WIRELESS COMMUNICATIONS IN DEVELOPING COUNTRIES

CELLULAR AND SATELLITE SYSTEMS

The Artech House Mobile Communications Series

John Walker, Series Editor

For a complete listing of the *Artech House Telecommunications Library,*
turn to the back of this book.

WIRELESS COMMUNICATIONS IN DEVELOPING COUNTRIES

CELLULAR AND SATELLITE SYSTEMS

Rachael E. Schwartz

Artech House
Boston • London

Library of Congress Cataloging-in-Publication Data
Schwartz, Rachael.
  Wireless communications in developing countries : cellular and satellite systems /
Rachael Schwartz.
    p.   cm.
  Includes bibliographical references and index.
  ISBN 0-89006-874-7 (alk. paper)
    1. Wireless communication systems—Economic aspects—Developing countries.
  2. Artificial satellites in telecommunication—Developing countries.   3. Cellular radio—
Developing countries.
    I. Title.
  HE9715.D48S38   1996
  384.5'09172'4—dc20   96-15595
                                                                   CIP

British Library Cataloguing in Publication Data
Schwartz, Rachael
  Wireless communications in developing countries : cellular and satellite systems
  1. Wireless communication systems—Developing countries 2. Artificial satellites in
  telecommunication—Developing countries
  I. Title
  384.5'091724

  ISBN 0-89006-874-7

Cover design by Darrell Judd

© 1996 ARTECH HOUSE, INC.
685 Canton Street
Norwood, MA 02062

International Standard Book Number: 0-89006-874-7
Library of Congress Catalog Card Number: 96-15595

10 9 8 7 6 5 4 3 2 1

*For my mother, Mary Jo Lakin, and my brother, John Lakin,
who would read anything I write.*

▼▼▼

Contents

▼▼▼

ACKNOWLEDGMENTS

This book arose from a paper that I wrote as a directed research project in connection with the International Development Clinic conducted by Professors Lea Brilmayer and Cathy Adcock at New York University School of Law. I am especially grateful to them for their permission to prepare that paper and for their comments.

Although many people and institutions provided documents and insights that were helpful to me in preparation of the initial paper, I wish to acknowledge the following who were particularly generous: Stephanie W. McCullough, International Trade Administration, United States Department of Commerce, who provided me with copies of the Argentine and South African requests for tenders as well as numerous articles, reports, and cables from her files; Mohammad Mustafa, World Bank, who arranged two days of interviews with World Bank experts and allowed me access to the Bank's Sector Library and other files; Ron Epstein, Volunteers in Technical Assistance, who was responsive to my telephoned questions; Martin Heath, KPMG-Peat Marwick, who sent me (and was a principal consultant on) the study on which the European Union's Green Paper on mobile communications was based; Mike Minges, International Telecommunications Union, who provided documents; and the Columbia Institute for Tele-Information at the Columbia University Business School, which opened its files to me.

Charlie Kennedy of Morrison & Foerster introduced me to his publisher, Artech House, and persuaded me that it would be worth the effort to expand the original paper to book length. Along the way he lent stout support. For all of this I thank him.

John Scott and Bill Wallace of Crowell & Moring generously agreed to review and comment on those parts of the book regarding FCC regulation. Any remaining errors are, of course, my responsibility.

Any opinions expressed in this book should be taken as those of the author, unless specifically indicated.

PART I

▼▼▼

OVERVIEW

CHAPTER 1

▼▼▼

INTRODUCTION

The past decade has seen enormous global growth in the use of wireless technology to satisfy the telecommunications needs of a variety of countries [1]. At year-end 1993, approximately 140 countries had cellular service [2]; even more have brought service online since then [3]. Satellites now make it possible to extend communications services to reach remote areas, although their cost has limited expansion of their use to date (as once was the case with cellular).

Cellular technology initially served solely as a means of mobile communications. It has been adopted by developing countries not only for that purpose, but also as a substitute, in both urban and rural areas, for a landline network [4]. (Industrialized countries, such as the United Kingdom, are now beginning to follow developing countries in this practice [5].) Satellites, too, are becoming a substitute for telephone wires in developing countries. And the most recent developments in the field of satellite technology will expand what has to date been the somewhat limited use of satellites for mobile communications. Satellites and cellular systems will both compete with each other and become integrated with each other in telecommunications networks.

This book is written to serve as a comparative study of how some countries, both developing and industrialized, have implemented wireless systems and some of the important policy decisions they have made in doing so. It discusses both mobile and fixed applications, focusing on two-way voice and data applications rather than on broadcasting. It is hoped that study of how various countries have put wireless solutions in place may be of some value to telecommunications regulators in developing countries that are contemplating such a course of action, to telecom-

---

3

munications lawyers who advise governments or investors, and to those investors themselves.

Unfortunately, given the great number of countries that use wireless technology and the limited space available to me, not every useful example can be examined here. However, I have made an effort to include among my case studies a diversity of countries in terms of geography, population, income levels, and philosophical approaches.

There are several ways to categorize countries in terms of stages of development. Two are the most common in the field of telecommunications. Material from the International Telecommunication Union (ITU) that lists the number of mobile subscribers in different countries classifies those nations according to income level. The ITU, headquartered in Geneva, Switzerland, operates in the United Nations system and oversees the world's communications resources [6]. Another classification scheme is to divide countries according to their "teledensity," or the number of telephone lines per one hundred persons [7].

To put this study of wireless in context, the book will first discuss the objectives that the government of a developing country may have in mind when it undertakes to modernize its communications system and how wireless in particular may satisfy these objectives. It will compare these goals to those of the investors whose participation governments frequently seek. To some extent, but not completely, the goals of government and investors are consistent. Indeed, even the goals of government involve trade-offs. For example, the greater the obligation to provide universal access to service that a government chooses to impose upon a cellular licensee, the less money it may be able to obtain in bids for the license. How the goals of government have been balanced among themselves and in relation to the goals of investors will be a key focus of this book.

A number of critical decisions will be examined. They include the following: Which technologies have been used and how have they been chosen? What services may be authorized over cellular or satellite frequencies? Is the operator to be a government instrumentality, a nongovernmental investor, or some combination? What restrictions on foreign investment have been imposed? How many licenses have been issued per country, for what term, for what coverage area, and by what processes? What regulatory structures—such as rate regulation, quality standards, and universal service requirements—have been applied to operators? A sample of some of the options that have been chosen to answer these questions will be presented.

I feel compelled to offer the following caveats. They may be obvious, but they are worth stating explicitly. First, an approach may succeed or fail in a given country for reasons that are endemic to that country. If the situation is not the same in the country seeking to learn from the experience, the results may differ. Second, adopting one aspect of a given country's approach may not bring about the same results in another country because the interaction with other parts of the approach will be missing.

Notes

[1]   For a thorough description of cellular technology, see D. M. Balston & R. C. V. Macario, *Cellular Radio Systems*, Norwood, MA: Artech House, 1993 [hereinafter "Balston & Macario"].

[2]   Jonathan R. Tarlin, "The Global Phenomenon: Global Cellular Communications Markets," *Cellular Marketing*, Feb. 1994, p. 18.

[3]   The United States Department of Commerce compiles lists of cellular providers (by country) and the number of their subscribers worldwide, and produces quarterly reports entitled "Summary of Recent International Cellular Developments."

[4]   For a helpful technical description of the use of radio-based technologies in rural areas not reached by landline systems, see D. Farrimond, "PCN and Other Radio Based Telecommunications Technologies for Rural Regions of the World," in *Second International Conference on Rural Telecommunications*, London: Institution of Electrical Engineers, 1990. For a discussion of the use of cellular by businesses in urban areas of developing and "transition" countries—such as Nigeria and Russia—where landline service is nominally available but in fact is very limited, see Vineeta Shetty, "Wireless Challenges Copper: Cellular vs Landline Telephone Services," *Communications International*, May 1993, p. 47. See also Greg Steinmetz, "Mannesman Scores with Cellular Service," *Wall. St. J.*, Feb. 24, 1995, p. A6 (describing use of cellular as a substitute for landline service in former East Germany).

    The term "landline" is used in this paper to refer to the traditional telephone company or its network, which has historically used physical lines, or wires, to connect subscribers. The term "wireline" will also be used in this book. Typically, the phrase "wireline cellular carrier" has been used to describe the cellular affiliate of a landline telephone company. This distinguishes it from the "nonwireline cellular carrier" in countries like the United States, where two cellular licenses were issued, one to a company affiliated with the landline company, and one to an unaffiliated company.

[5]   "Venture to Push Wireless Phones in British Homes," *Wall St. J.*, Nov. 2, 1995, p. B8.

[6]   For a discussion of the ITU and its history, see Audrey L. Allison, "Meeting the Challenges of Change: The Reform of the International Telecommunication Union," 45 *Fed. Communications L. J.* 491, (1993). See also Ch. 16, *infra*.

[7]   *E.g.*, Peter L. Smith and Gregory Staple, "Telecommunications Sector Reform in Asia: Toward a New Pragmatism," *World Bank Discussion Paper No. 232*, p. 92 (1994) (low teledensity countries have fewer than one main line per 100 inhabitants, medium-teledensity countries have 1–20, and high-teledensity countries have more than 20)[hereinafter "Smith and Staple"]. Differently calibrated teledensity scales have also been used. *E.g.*, Andrew Adonis, "Survey of IMF World Economy and Finance," *Fin'l Times*, Sept. 24, 1993, p. xxv (developing countries have fewer than 10 lines per 100 people, emerging market countries have between 10 and 30, and developed countries have more than 30).

CHAPTER 2
▼▼▼

GOALS OF GOVERNMENT
AND INVESTORS

2.1 GOVERNMENT GOALS (PUBLIC BENEFITS)

The benefits that a strong communications infrastructure, accessible to all, can yield to the citizens of developing countries are numerous [1]. It provides a lifeline, especially for those in rural areas. It facilitates more effective internal operation of the country's markets, government, and private social service organizations. It is a link to the rest of the world that is necessary to attract foreign investment and tourism and to draw on the information resources physically maintained in other countries. Another important, but less tangible benefit is a decreased sense of isolation. Finally, a telecommunications system can serve as a source of revenue to the government: revenue that can fund the provision of other important services to the public.

Medical and other emergency services that are not typically close at hand in rural areas of developing countries can be summoned or consulted by telephone. Doctors can be called by phone, and even if they cannot respond in person they can give useful advice in situations where time is at a premium (e.g., in the case of selecting the correct serum for poisonous snake bites and treating other mortal wounds). Audio teleconferences can be made a part of the training of health care workers. When natural disasters strike, assistance can be summoned. Drivers carrying important supplies can call for help in the event of breakdown [2].

Given that markets function best when information is available to all participants, the value of a well-functioning telecommunications system to economic efficiency is clear. Farmers have been able to obtain better prices for their crops because they have had telephone access to information on prices in urban areas; the end result can be an ability to expand farm production because of the additional funds [3]. Companies have received more orders for their goods because customers have been able to contact them by phone; this allows companies to hire more employees. When opportunities in less populated areas increase, urban overcrowding can be reduced or avoided entirely [4].

Government agencies can function more effectively when telecommunications services are available. They can substitute telephone calls for time-consuming travel in order to hold meetings and otherwise exchange information and ideas [5]. Laws can be enforced when officials can be beckoned to perform this function. (Conversely, one might contend, with some force, that sophisticated telecommunications systems can help criminals avoid detection by law enforcement authorities [6].)

Other social service organizations can also use telecommunications to be more productive. An example is the efforts of nongovernmental organizations (NGOs) to increase respect for human rights and improve the provision of basic services. Better communication among NGOs and between NGOs and their constituencies can increase their effectiveness [7]. Of course, it is for exactly this reason that a repressive government would not want a functioning telecommunications system to be available to anyone other than itself [8].

It is difficult to attract foreign investment without a telecommunications system that allows investors to maintain contact with home and other branch offices. And not all tourists wish to travel in areas in which they will be completely isolated from the outside world while on vacation.

The ability to make contacts outside the country can help in the situations described above and more. Many resources (e.g., data banks, expertise) are physically located outside of any given country. Availability of telecommunications can also serve to enhance quality of life in less concrete ways. These include the ability to contact family and friends and a sense of national and global cohesion. Moreover, when adequate communications facilities are lacking, the likelihood of misunderstanding (such as political problems between countries that can lead to violence) is magnified.

The government's financial well-being can also be enhanced by implementation of telecommunications systems. The franchise to provide services can be sold by the government for an up-front fee and/or a share of the business' revenues or profits. Government may become a minority investor in the entity providing the service. Alternatively, the business may be operated at a profit by the government itself. And much of the increased financial activity described above can lead to increased tax revenues.

Wireless is particularly well-suited to achieving many of these goals, for a variety of reasons. Cellular is a mobile technology. It can be less expensive than a landline system in some configurations. It discriminates less among those served because anyone has the physical ability to obtain service when within reach of a cell site (although there may still be discrimination in the placement of cell sites). It can be installed more quickly than a landline system and it can be particularly profitable. The user's ability to be "untethered" allows cellular to be used not only in the same situations as a landline phone, but also from places (such as an automobile) where landline service is physically impossible [9]. The absence of a need to install and maintain lines over long distances to serve remote areas is an important factor in keeping the cost of cellular below that of a landline system [10]. And one need not choose between waiting for a line to be installed to one's home or office and paying an exorbitant amount of money to obtain a connection quickly in order to procure phone service. Rather, one can quickly subscribe to cellular service or take advantage of cellular pay phones [11]. A cellular system can often be installed in relatively short order [12]. And the sale of a cellular franchise can be quite lucrative for a government [13].

Much of this is also true of satellite-based systems, although many of them are still being developed and are or will be more expensive than cellular systems. The primary advantage of satellites over cellular is the broader coverage (or bigger "footprint") they afford. The possible drawback is that a developing country will have less regulatory control over the provision of satellite communications and a satellite system would probably provide less revenue than could be derived from nationally provided service [14]. Nevertheless, if the price of satellite service declines, one can expect the degree of substitutability between cellular systems and satellite systems to increase.

2.2 INVESTOR GOALS

Achieving the governmental goals discussed above, however, often calls for the involvement of outside investors to build and operate the communications system. Such investors have capital and expertise that are quite valuable in such a venture. To a degree, the goals of investors overlap with those of government. For example, providing high-quality service will usually be in the interests of both government (for all the reasons discussed above) and investors (high-quality service will attract more customers). However, the congruence of public and capitalist interests must often be facilitated through conscious structuring of incentives. To illustrate, the government may want to award a cellular license to a potential operator who offers a large upfront payment for the license in the hope that this will encourage the operator to build its system as quickly as possible so as to recover in profits the up-front payment. However, if the up-front payment required is too large, potential operators may be discouraged from bidding at all.

Assuming that outside investment is desirable from the government's point of view, how does it attract such investment? There is a limited amount of capital available, and countries are in competition with each other for it [15]. How do investors evaluate the various opportunities available to them?

An example of how cellular investments are evaluated is presented by the case of the investment by the International Financial Corporation (IFC), an affiliate of the World Bank, in Clovergem Celtel Limited (CCL). The International Bank for Reconstruction and Development (IBRD), also known as the World Bank, and its affiliate, the IFC, provide development assistance. The IBRD was originally formed to help war-torn European countries rebuild their economies. The IBRD may make direct loans to governments out of its own funds and from funds it raises in financial markets; it may also guarantee loans made by private investors. The IFC "is closer to being a private international investment organization than is the World Bank and can assist states with special sorts of financial assistance such as contributions to equity and underwriting capacity" [16]. The World Bank may make, participate in, or guarantee loans to the IFC. Both the IBRD and the IFC are involved in international cellular issues in developing countries. The IFC makes investments in cellular operators. It also advises governments on issuing invitations to investors who wish to tender bids for cellular authorizations. The IBRD advises governments on privatizing the telecommunications sector (including but not limited to wireless) and opening it to competition [17].

The IFC decided in the summer of 1994 to make loans to and an equity investment in CCL, which was applying for a national cellular license in Uganda. The IFC would own 10.5% of CCL. In deciding to make its investment, the IFC performed a financial analysis of the cost of the project [18], estimated its rate of return, and made projections through the year 2002 of revenue, net income, cash flow from operations, total assets, long-term debt, shareholders' equity, current ratio [19], long-term debt/(long-term debt + equity), and long-term debt service coverage. It noted that the landline system was underdeveloped due to a civil war and had only 23,000 lines serving a population of 17 million (with 3,000 of those lines out of service and call completion rates below 50%) and a waiting list of 31,400 persons. The IFC estimated the number of subscribers it would have over the next 5½ years, forecasted expected usage per customer, and assumed prices competitive with those in other African countries (which prices would decline over time). However, it also evaluated qualitative factors. It considered the identity and experience of other investors in and the managers of the system, the nature of the cellular technology authorized and whether it was compatible with other technologies being used in neighboring countries, and the fact that CCL had agreed to comply with World Bank environmental and safety guidelines. Also important was Uganda's regulatory structure: that the government had sought World Bank assistance in privatizing the telecommunications sector; that the license was valid for 15 years, renewable for 10 years (subject to review every 5 years to ensure fulfillment of license terms and conditions); that the network was required to be operational within 2 years (a longer time than is often the case); that CCL would be free to set its own prices; and that while the license was not exclusive, the Uganda Posts and Telecommunications Corporation had agreed not to allow another cellular provider to interconnect with the landline network for 2 years [20]. It considered the ancillary economic benefits to other sectors of the economy and the number of jobs to be generated by the project. In terms of risk, the IFC noted that the many uncertainties of the Ugandan market were mitigated by such factors as the pent-up demand for phone service; the vibrant informal economy; various externalities (including security), which are likely to drive up demand; and the known market in business, government, and international organizations and the expatriate community. Also significant was the ability of CCL to convert its profits from Ugandan to foreign currency following the liberalization of foreign exchange controls in November 1993 [21].

A purely private investor, such as a multinational cellular company, would not be constrained by World Bank policies and thus would evaluate a prospective investment slightly differently. It might, nevertheless, be influenced by whether the country issuing the license had sought World Bank assistance in privatizing its landline telephone provider if it felt that this indicated an attitude receptive to private investment [22]. Whether a result of World Bank encouragement or otherwise, liberalization of telecommunications policy is often cited as a drawing card for governments seeking to attract private investment in the area [23]. But it would seem unlikely to take into account, for example, whether the country at issue had agreed to be bound by World Bank safety and environmental guidelines.

A private cellular company, in turn, would be interested in facts that the World Bank might find irrelevant. For example, such companies are usually invited to bid on licenses because their employees possess cellular industry expertise that would be valuable to a developing country. Therefore, a private company must concern itself with whether the country seeking bids is one for which it could recruit talented employees to live in for at least several months, if not a matter of years. The personal safety and quality of life of the employee are therefore important issues. Moreover, the company must consider the substantial costs of supporting expatriate employees, which are significantly in excess of compensating the same employees in their home country [24].

However, there is also likely to be a great deal of overlap between other qualitative considerations and most quantitative factors considered by the IFC and those relevant to the decision of a purely private investor. As was the IFC in the case of Uganda, a multinational corporation would also be concerned about the political stability of the country. And private investors prepare quantitative analyses that are quite similar to (if sometimes more detailed than) the World Bank analysis discussed above [25]. Often, in fact, such detailed projections are required to be included as part of the investor's bid package.

As noted, two factors that sometimes stand in the way of telecommunications development are scarcity of capital and political risk. The World Bank, as discussed, can alleviate some of the difficulty of obtaining capital for investments in developing countries. Some additional relief from these problems can be provided by agencies such as the Overseas Private Investment Corporation (OPIC), the European Bank for Reconstruction and Development (EBRD), and other similar agencies.

OPIC is a U.S. government agency that encourages American businesses to invest in developing countries and emerging market economies. First, it finances investments through medium- and long-term direct loans and loan guarantees. Second, it provides political risk insurance against expropriation, political violence, and inability to convert foreign currency into U.S. dollars. OPIC currently operates in 144 countries worldwide [26]. As one commentator has noted, a particular advantage of utilizing OPIC as a lender "is that the host government tends to recognize that it may have more to lose than gain by interfering with the private business interests involved" [27].

OPIC has provided assistance in telecommunications projects, including wireless. It provided 200 million dollars in financing for a cellular network to be installed by GTE Mobile Communications International and AT&T International in Argentina. It has been especially active in Russia: it provided political risk insurance for a small U.S. company creating a wireless network to bypass the landline system and to another American company for the installation of satellite uplink stations. It also granted a 125 million dollar loan guarantee to U S West, a regional Bell company, to fund telecom projects in Russia, including the construction of mobile digital cellular networks in 12 cities [28].

Political risk insurance is also available from other sources. Private insurers offer this product [29]. And, in addition to its IBRD and IFC branches, the World Bank has another arm, the Multilateral Investment Guarantee Agency (MIGA), that makes political risk insurance available to eligible investors from member nations for investments in a developing country that is also a member [30].

Loans, loan guarantees, and political risk insurance are available from a number of other development agencies. The European Bank for Reconstruction and Development (EBRD) offers them for projects in the former Soviet Union and in Eastern and Central Europe. The Inter-American Development Bank (IDB) finances projects in Latin America, including making loans to the private sector and issuing loan guarantees. The Asian Development Bank (ADB) lends to private sector entities for projects in its regional member countries where the goods and services are procured from member countries. (ADB has nonregional members, including the United States and several from Europe.) The African Development Bank serves a similar function for its African regional member countries. (It too has nonregional members.) The European Investment Bank (EIB), the financial institution of the European Union, provides financing to private sector borrowers for projects in the less developed countries of the EU and in certain African, Caribbean, and Pacific countries that are signatories of the Lomé Convention and certain Mediterranean, Eastern and Central European, Asian, and Latin American countries that have signed agreements with the EU [31].

2.3 COMPARISON OF CELLULAR AND SATELLITE INVESTMENTS

The expectation in the telecommunications industry has been that as the costs of satellite systems decrease, satellite service will become competitive with cellular service. Therefore, it is worthwhile to examine briefly some of the differences, from a cost point of view, between cellular and satellite systems.

The costs of a satellite system are primarily "upfront costs." These include building and launching the system. A satellite system, in its early years, may have much more capacity than it needs. In contrast, a cellular system is more "modular." As the coverage area and number of subscribers increase, the capacity of the cellular system can be expanded through additional construction.

Conversely, satellite costs are constant with respect to geographical area up to the limits of the coverage area (footprint). For very large territories, satellite investment is therefore more efficient [32].

Cellular has greater flexibility to conform its investment patterns to population concentrations. Thus, the costs of a cellular system can be reduced if it is built to serve minimal, high-density areas (e.g., a busy highway). The downside is that if a subscriber travels into an unserved area, the system becomes useless. The same is true for those who live in unserved areas and depend on a fixed terrestrial wireless system for their basic telecommunications needs [33].

2.4 CONCLUSION

A government that wishes to attract private investment in a wireless system will want to understand how a private investor—such as an international telecommunications company—or analogous entity—such as the IFC—evaluates opportunities for investment. A developing country may or may not choose to implement certain policy changes or incorporate certain features of a regulatory scheme into its own format in order to attract profit-maximizing investors. For example, cellular operators are looking for stable political systems, something that is in the government's interest to establish as well. But, for example, a cellular company may wish to use least cost distribution systems (such as its own employees), whereas the government may want the company to use local enterprises as agents in order to encourage local entrepeneurship, even though this may involve additional expense to the operator. Knowing what private investors are looking for, however, will help in several ways. First, as noted, it may affect basic government policies or the structure of the investment opportunity being offered. Second, it will enable the developing country to predict what sort of bids it can expect from prospective licensees. Third, it will create an awareness of what may be considered to be the positive aspects of the country, which can then be brought to the attention of potential investors.

Conversely, investors would do well to understand the needs of the countries in which they seek to provide service. Such an understanding is of obvious benefit in formulating bids. However, it is also useful in the pre-bid stage, when investors may be permitted to lobby the government or submit comments regarding draft tenders issued by the government.

Notes

[1]   Much of this discussion of the benefits of improved telecommunications is (except where otherwise indicated) taken from examples contained in Robert J. Saunders, Jeremy J. Warford, and Björn Wellenius, *Telecommunications and Economic Development*, 2d ed., The Johns Hopkins University Press, 1994, pp. 15–29 [hereinafter "Saunders"]. The authors of that book, in addition to setting forth anecdotal evidence of benefit, examine the statistical evidence in an effort to quantify this benefit. As might be expected, they find this to be a difficult task at which they admit they only partially succeed. *Id.* at 83-195.

[2]   *E.g.,* Carol S. Saunders, "Solar Phone: Drivers' Friend in Need," *N.Y. Times*, Jan. 8, 1995, § 13, p. 1, col. 4.

[3]   Saunders, *supra* note 1, at 23.

[4]   D. Westendoerpf, "Development of Rural Telecommunications and the CTD," *in Second International Conference on Rural Telecommunications*, London: Institution of Electrical Engineers, 1990.

[5]   Travel reduction can also result in decreased pollution and increased labor potential.

[6]   Cellular phone conversations can be more difficult than landline calls for government to eavesdrop on. A recently passed U.S. law requires cellular carriers, among others, to redesign their networks to make them more amenable to wiretapping. See *Communications Daily*, Nov. 7, 1994, p. 5 (reporting on President Clinton's signing of the legislation); Elizabeth Weise, "Feds log on with bid to conduct surveillance," *S.F. Examiner*, Oct. 17, 1994, p. C-1; "Congress Passes Wiretap Legislation; Senate Approves by Unanimous Consent," *BNA Washington Insider*, Oct. 12, 1994. See also "Postscripts," *Wall St. J.*, May 3, 1995, p. A11 (describing draft bill in Germany to force cellular operators to allow police to monitor calls in combating organized crime).

[7]   Timothy A. Johnston, *Technology and Electronic Communications for African NGOs*, May 1993 (unpublished paper on file with the author). See Philip Shenon, "A Repressed World Says, 'Beam Me Up,'" *N.Y. Times*, Sept. 11, 1994, Editorial Section, p. 4.

[8]   *E.g.,* "Difficult Market Conditions; CIS Seeks Economic Support from World Bank, IMF and Joint Ventures," *Comm. Daily*, April 27, 1992, p. 3 (reporting comments of World Bank to the effect that "[u]nder communism, telecommunications was given low priority—officials were afraid of giving public unrestricted access to international information flows, which they feared would 'encourage democratic forces,'"); Anthony Ramirez, "Cellular Phones Fill Gap in Hungary," *N.Y. Times*, July 6, 1992, § D, p. 2, col. 4 ("As in the former Soviet Union, the Communist party wanted to keep phone service scarce and primitive to facilitate eavesdropping while crimping the flow of communications among citizens.").

[9]   *E.g.,* Saunders, *supra* note 1, at 25, 138.

[10] Westendoerpf, *supra* note 4, at 18–19.

[11] "The Third World is Getting Cellular Fever," *Business Week*, April 16, 1990, p. 80 (describing bribe demanded by employee of Télefonos de México to rig an illegal connection for a businessman who had waited four years for a phone); "Fighting for Franchises Where No Phone Has Gone Before," *Business Week*, April 16, 1990, p. 81 (reporting on solar-powered cellular pay phones in remote villages in Borneo).

[12] *E.g.,* "RHC Strategies Discussed; Foreign Cellular Growth Seen Mushrooming, But Doubts Raised on CT-2," *Comm. Daily,* Jan. 22, 1991, p. 4 (a Bell South consortium was able to put a cellular system in Guadalajara, Mexico in service in less than 60 days).

[13] *E.g.,* "Colombian Cellular Licenses Awarded," *Latin Am. Telecom Rep't,* Feb. 15, 1994, p. 7 (Colombian government to receive approximately $1.186 billion from sale of cellular franchises).

[14] The development and expense of satellites is discussed in Jeff Cole, "Hughes Lands India-Based Contract for Hand-Held Mobile-Phone Service," *Wall St. J.,* Jan. 23 1995, p. A2, col. 2; John J. Keller, "TRW, Teleglobe Set $2.5 Billion Project for Satellite and Wireless Phone System," *Wall St. J.,* Nov. 14, 1994, p. A2; Richard L. Hudson, "Inmarsat Begins Fund-Raising Drive for $2.5 Billion Satellite Phone System," *Wall St. J.,* Sept. 12, 1994, p. B8; Balston & Macario, *supra* Ch. 1, note 1, at 345–50; Saunders, *supra* note 1, at 60 n. 19. The coverage advantage of satellites is addressed in *ITU Africa Forum; Mobile and Fixed Satellite Service Operators.* See "Opportunities in Africa," *Comm. Daily,* May 16, 1994, p. 3; Saunders, *supra* note 1, at 60 n. 19. See also Mary Lu Carnevale, "FCC, Clinton Administration Both Move to Spur Satellite Network Competition," *Wall St. J.,* Nov. 18, 1994, p. B4. The ITU noted the issues relating to governmental control and revenue in *International Telecommunications Union,* "World Telecommunication Development Report 1994" at 348.

[15] *E.g.,* Michael R. Sesit, "Dollar Darwinism: Global Capital Crunch is Beginning to Punish Some Weak Economies," *Wall St. J.,* Jan 12, 1995, p. A1, col. 6.

[16] Mark Janis, *An Introduction to International Law,* Little, Brown and Co., 1992, pp. 215–16.

[17] Lisa Sedelnik, "Getting in Step; Latin American governments turn to private sector for infrastructure needs," *Latin American Finance,* June 1994, p. 22; Conversation with Peter Smith, Senior Telecommunications Policy Specialist at the IBRD, Sept. 28, 1994.

[18] The costs were infrastructure, network control system, building and civil works, office equipment and furniture, working capital, interest during construction, and contingencies. No fee was payable up front; rather, CCL was obligated to pay a fee calculated on a per-channel per-cell basis.

[19] "The ratio computed by dividing current assets by current liabilities; indicates the extent to which the claims of short-term creditors are covered by assets expected to be converted to cash in the near future." Eugene F. Brigham, *Fundamentals of Financial Management,* 5th ed., The Dryden Press, 1989, p. 267.

[20] See Ch. 13, *infra* for a discussion of the significance of landline interconnection.

[21] Discussion with Mohammad Mustafa, Senior Financial Analyst, The World Bank, Sept. 28, 1994.

[22] Telephone conversation with Mohammad Mustafa, Jan. 10, 1995.

[23] *E.g.,* "Sprint runs to Israel; bids in contract for cellular telephone channel," *Israel Business Today,* Feb. 25, 1994, p. 5.

[24] Conversation with Robert VanBrunt, Vice President, Wireless Development, Bell Atlantic International Wireless, Nov. 3, 1994.

[25] For example, such analyses might contain analyses of distribution costs by type of channel, breakdown of projected customer usage by price plans or time of day, and expected churn rates (the percentage of customers that will disconnect each month). The projected financial data might be varied to reflect different discount rates.

[26] Testimony July 13, 1995, Christopher Finn, Executive Vice President OPIC, before the Senate Foreign Relations Committee, Near Eastern and South Asian Affairs Subcommittee, on Economic Development in Jericho and Gaza; Testimony March 28, 1995, Ruth R. Harkin, President and Chief Executive Officer OPIC, before the House Appropriations Committee, Foreign Operations, Export Financing and Related Programs Subcommittee, on FY96 Foreign Operations Appropriations.

[27] Howard L. Hills, "OPIC: A Government-Business Partnership That Works," *Legal Times,* June 20, 1994, p. 42.

[28] Testimony of Ruth Harkin, *supra* note 26; "OPIC Grants Largest-Ever Loan Guarantee to US West," *Global Telecom Rep't,* April 18, 1994.

[29] For a description of political risk insurance and private and public entities that provide it, see Cecil Hunt, "Political Risk Insurance: Government and Private Market," in *Negotiating and Structuring International Business Transactions*, American Conference Institute, 1995.

[30] "OPIC: A Government-Business Partnership That Works," *supra* note 27; Cecile Gutscher, "Political Risk Insurance in Booming Third World Business," *LDC Debt Rep't/Latin American Markets*, Jan. 24, 1994, p. 1.

[31] Roger Weigley, "Financing International Projects," in *Negotiating and Structuring International Business Transactions*, American Conference Institute 1995).

[32] Steven Adamson, "Advanced Satellite Communications: Potential Markets," Noyes Data Corp., 1995, pp. 311–12.

[33] *Id.* at 313.

PART II
▼▼▼

CELLULAR

CHAPTER 3

▼▼▼

TECHNOLOGY

3.1 INTRODUCTION

Cellular technology employs a grid of hexagons, or cells, that cover specific geographic areas. Each cell contains a low-powered radio transmitter and control equipment located in a building called a *cell site*. (Outside the United States, cell sites are often referred to as *base stations*.) The cell site is connected by wireline, microwave, or satellite facilities to a mobile telephone switching office (MTSO), which is typically then connected to the regular landline network through the local telephone company's central office. (Direct connections from the MTSO to long-distance carriers can also be made, bypassing the local telephone company's central office.) With its electronic switching capability, the MTSO monitors the signal strength of the mobile units used by customers and automatically *hands off* conversations in progress to the next cell site as the customer moves from one cell to another. Each cell has a set of radio frequencies, allowing reuse of every channel for many different simultaneous conversations in a given service area. As demand for the service grows, dividing cells into smaller cells allows greater numbers of customers to be served.

Thus, cellular is a wireless technology in the sense that the piece of equipment used by the customer (the phone) communicates with the network through radio waves rather than through a wired connection. This fact is significant with respect to technology for at least two important reasons. First, there are numerous transmission standards that define the air interface, or the link between the customer's equipment and the cell-site portion of the network. To analogize, the phone and the cell site must speak the same language to understand the signals they send to each other.

Because there are a number of air interface standards in use throughout the world, either the government or the marketplace must elect which one(s) to adopt. Second, the wireless aspect of cellular means that it can accommodate both mobile and fixed uses. Depending on whether a country wishes its network to serve one or both types of these functions, different network equipment and configurations are available, at different costs.

3.2   AIR INTERFACE TECHNOLOGIES

The choice of a particular air interface is important for several reasons. Any country that is considering issuing a cellular franchise will want to use a technology that will be reliable and have sufficient capacity to serve its citizens well, while not being unduly expensive to implement. But the choice is significant for another reason: there is an advantage to making it as easy as possible for customers visiting from other countries to use cellular service—for example, by allowing them to bring their phones with them [1]. It is from these customers, referred to as *roamers*, that a substantial amount of the revenue of a cellular business may be derived. Typically, roamers are charged higher rates than *home* customers. Roaming charges may include a daily fee as well as a higher per minute rate. Therefore, a country would want a technology that serves not only its own citizens, but its foreign visitors as well. Conversely, its choice might also be affected by the desire for its citizens to be able to use their phones when they are abroad. (Even given the portability of phones and the identity of the air interface standard, however, a customer may not be able to use the same phone she uses in her home system in a foreign country, because the foreign system may operate at a different frequency; that is, on a different part of the radio spectrum.)

The first distinction to be made between different types of transmission technologies is the one between analog and digital [2]. Originally, all cellular systems used analog radio signals, which mimic the wave pattern of the human voice to convey conversations and data. However, digital technologies, which can fit more transmissions into the same amount of spectrum, have since been developed. Digital radio signals transmit by converting the speech or data being sent to the 0's and 1's of computer language [3]. The increased capacity of a digital system may be of particular interest in developing countries, which often have very crowded urban areas and thus a high concentration of phone users in a small area. However, even digital technologies may not provide sufficient capacity to solve this problem in some countries [4]. In addition to using the airwaves more efficiently, digital technology may make it more difficult for eavesdroppers to intercept telephone calls [5].

Analog is not without its advantages, however. It may be deployed more rapidly than digital, it is cheaper, it can potentially provide a greater coverage area from a given cell site, and it is subject to fewer export restrictions [6]. One other advantage to a developing country of using an analog standard is that used analog equipment may be available through service providers who are switching to digital and who might be able to resell their equipment in other countries.

The major analog standards in place are Advanced Mobile Phone System (AMPS), Nordic Mobile Telephone (NMT), and Total Access Communications System (TACS) [7]. AMPS is dominant in the Americas, but has spread to other parts of the world. A more advanced form of AMPS technology, Narrowband AMPS (N-AMPS), allows for increased system capacity by splitting each 30-kHz AMPS channel into three 10-kHz channels [8].

NMT, as the name implies, originated in Scandinavia but is also in use in many other European countries (particularly eastern Europe), some states of the CIS (the former Soviet Union), and African countries such as Algeria, Morocco, and Tunisia. NMT-450 refers to equipment that operates in the 450-MHz frequency range of the spectrum, and NMT-900 to that which operates in the 900-MHz range. A system that begins on the 450-MHz range can later be expanded to include the 900-MHz range [9].

TACS originated in the United Kingdom, but is now used as well in other parts of Europe and in Asia. Two more advanced versions of TACS are ETACS and TACS 2. ETACS incorporates modifications that increase the spectrum allocation (and hence the number of channels) and TACS 2 includes improvements that enhance the signaling and reduce fraud. NTACS expands the original capacity of a TACS system by narrowing channel separation. JTACS, which is based on the U.K.'s TACS system, was put into service in Japan in 1989 [10].

There are two basic types of digital technologies currently contending in the cellular arena: time division multiple access (TDMA) and code division multiple access (CDMA). Analog technology sends one conversation over one channel. For this reason the technology used by analog systems is called frequency division multiple access (FDMA). The FDMA approach can also be used with digital, but that would entail forgoing the spectral efficiency gains associated with digital [11]. In contrast, TDMA breaks a number of conversations into fragments that it transmits over one channel at precisely timed intervals [12]. CDMA also fragments conversations, but does not rely on timing to determine how to reassemble the pieces at the receiving end. Rather, it attaches a unique identifier, or code, to each piece. TDMA is expected to at least triple capacity compared to analog. The developer of CDMA at one time claimed improvements in capacity on the order of 18 times analog. The estimate of capacity expansion for CDMA has since been reduced to 10 times that of analog [13].

TDMA technology is used in two important digital standards: GSM (originally derived from Groupe Speciale Mobile, [14] but now modestly called Global System for Mobile) and Interim Standard (IS)-54 [15]. GSM has been adopted by the European Union as a mandatory standard for member states and is spreading throughout much of the world [16]. However, the United States and other countries that have significant investments in their existing analog AMPS cellular systems (or another analog technology) have chosen to transition the networks they have in place to digital over time: some channels will use analog technology and some will use digital.

The proportion of digital to analog channels in the network will increase over time. This means that the large number of customers who already have analog phones will not have to replace them immediately and simultaneously. Dual-mode phones, that use both analog and digital technology, will also be available [17]. At the same time, this solution allows the carriers to resolve their capacity problems. To accomplish these goals, the digital technology these countries use must be capable of being over-laid on the existing analog network. IS-54 TDMA technology was developed for this purpose. CDMA technology was also developed to address this problem. When GSM was designed, no attempt was made to make it compatible with the various analog networks already in place in Europe, the Americas, or any other part of the world [18].

Even though cellular licensees in the United States are not using GSM, some of the new operators of personal communications networks that received licenses beginning in 1995 will be [19]. (The new licenses for personal communications networks are discussed in more detail in Chapter 7.) Because these companies are building new networks and do not have an embedded base of customers with analog phones, they can begin with GSM technology.

The proliferation of GSM systems worldwide can make it attractive to developing nations [20]. A traveler may take his or her phone to another country that uses a system that is like the one used at home. Additional flexibility is provided by the fact that the GSM system operates by allowing any user to have a subscriber identification module (SIM) card. The user inserts the SIM card into the phone's handset in order to activate the phone and verify the customer's identity and authorization to use the network [21]; the customer may also be required to enter a personal identification number [22]. The SIM card allows a customer to use a phone that she or he does not own, such as a rental phone or borrowed phone, and still have the charges billed to her or him directly. Even when a roaming customer cannot use the cellular phone that she or he uses on her or his home system in a foreign country (because, for example, the phone is not portable but is installed in the car), she or he can borrow or rent a phone in the foreign country and make calls if it operates on a GSM system that will allow the use of a SIM card. This is likely to be less expensive than buying the service from a middleman (such as the lessor of the phone). Some cellular carriers that provide service on AMPS systems in the United States have made it easier for their customers to obtain GSM service while they are traveling in Europe, and in some cases vice versa [23]. The operators of the home and visited networks must have entered into a roaming agreement allowing them to bill each other for usage by each other's customers. (This is true even when the systems use the same air interface technology.)

Another advantage of the wide development of GSM digital cellular systems is that it has led to large-scale production of phone sets at reduced prices. Although the end user rather than the government bears this expense, high-priced handsets would certainly pose an obstacle to the government's goal of deriving social benefit from increased access to telecommunications [24].

In addition to achieving widespread acceptance in Europe and a measure of success in Asia, GSM is also being adopted by some African and Middle Eastern countries as the standard for one or more of their cellular franchises. These include Egypt, Algeria, Morocco, Tunisia, Cameroon, and South Africa [25].

However, GSM has been criticized on the ground that coverage area per cell site is too small in comparison with CDMA. This can increase the cost to such a degree that GSM may not be viable for some developing countries [26].

3.3 MOBILE, FIXED, AND INTEGRATED USES

3.3.1 Introduction

In most parts of the industrialized countries, cellular technology was initially developed to serve mobile users traveling in automobiles. In rural parts of these countries, and in developing or transition countries that lack a strong landline infrastructure, wireless technology—including but not limited to cellular technology—has been used as a way to serve homes and offices [27]. This is a trend often referred to as *wireless in the local loop*. Over time, however, mobile and fixed uses have converged so that one system can serve both classes of customers [28]. The convergence of fixed and mobile aspects of a network is sometimes referred to as personal communications services (PCS) or the personal communications network (PCN). This section will discuss wireless in the local loop and PCS/PCN.

The use of wireless in the local loop has increased as the cost associated with it has decreased more quickly than the cost of a landline system over the years [29]. Cellular as a substitute for landline, or wired, service has the particular advantage of being *modular*. That is, the initial investment to erect cell sites can be made and additional equipment can be purchased as more customers sign up for service. In this way, expenditures for equipment beyond the initial build-out can be made out of the revenues of the system. (And the initial build-out itself may be financed by the vendor of the network equipment [30].) Building a landline system cannot follow this pattern; one does not want to redig trenches to lay a second cable [31]. A cellular network built to accommodate both fixed and mobile uses may well be more expensive than other types of wireless networks that are designed solely to serve fixed uses (such as fixed point-to-multipoint radio systems). Employing a system that serves only fixed uses eliminates the need for a considerable amount of switching and software associated with mobile service [32]. However, a cellular system designed to serve both fixed and mobile uses presents a unique opportunity not available with less expensive systems engineered solely for fixed uses: the chance to subsidize service to fixed subscribers who are not well off with revenues from affluent mobile users [33].

PCS/PCN is a term that has been used in several ways and that can cause much confusion if the speaker's intent is not clear. This chapter will discuss three meanings. The first is the evolution of existing cellular systems from systems that were intended to serve mobile users in vehicles to systems that will deliver voice and data communications to anyone, anytime, anywhere. The second is a cordless voice telephone system that may not be able to receive incoming calls and that is not suitable for people traveling in cars, because it lacks full ability to hand off calls from cell to cell. The third is a system that may look much like what the existing cellular system is evolving into, but that will be provided at a higher frequency than existing cellular.

3.3.2 Evolution of Cellular Technology

Cellular technology was originally developed for use in cars [34]. Therefore, cells were designed to cover areas where people travel in cars. Radio waves from these *macrocells* may not penetrate deeply into buildings and may be blocked on a street surrounded by skyscrapers. Through the use of *microcells* and *picocells* [35] to fill in the coverage gaps, cellular is evolving into a system that can also be used by callers who either do not move or who walk rather than drive. (The main purpose of microcells and picocells is to allow coverage indoors. However, they may also be used as temporary outdoor cell sites in emergency situations, such as natural disasters.) Moreover, by utilizing advanced intelligent network capabilities built into the landline network, calls can be delivered to an individual whether he or she is in the car, office, home, or is walking down the street. The hope is that eventually the intelligent network function will be built into the customer's cellular equipment. At home or office, the handset can be plugged into a base station and used like a cordless phone.

The United States Federal Communications Commission has recently proposed that broadband commercial mobile radio service providers (a category that includes cellular carriers) be explicitly authorized to provide fixed wireless local loop services. The Commission stated that it was taking this step to foster competitive local exchange service by allowing wireless carriers to provide the equivalent of local exchange service [36].

Cellular technology was also originally developed for voice communications [37]. However, technological evolution has led to cellular networks being used for data as well. For example, cellular carriers are developing and using cellular digital packet data (CDPD) technology for both mobile (e.g., cellular fax, e-mail, and vehicle location) and fixed (e.g., cellular alarm systems and utility monitoring) data applications. CDPD is designed to be a frequency-agile technology that breaks data transmissions into bursts, or packets, that can be inserted into the blank spaces in conversations on different channels and then reassembled at the receiving end. Data can also be provided on a circuit-switched basis, in which one channel is dedicated to the transmission (as it would be for a conversation), but CDPD will be more efficient because it will not tie up channels that could otherwise be used for voice transmissions. Even when a single channel is devoted to CDPD, more than one transmission can be carried on it. However, circuit-switched technology is better suited to certain applications that involve long, continuous transmissions (e.g., file transfer of a word processing document) than is packet data technology [38].

3.3.3 Cordless Telephony

A second form of PCS/PCN developed along a path that split off early in the process of cellular evolution. For this reason, as well as its relative lack of success in some areas, it might be referred to as the Neanderthal family of PCS/PCN. It is called CT-2 and represents the second generation of cordless telephones. Like the first generation of cordless phones, the subscriber equipment comprises two parts: a base station and a handset. However, the CT-2 handset is lighter and smaller and thus portable.

CT-2 has certain technical characteristics that make it less expensive than traditional cellular. Capability for handoff of ongoing conversations when the caller moves from the coverage area of one cell to the next is limited. Part of the problem is that the network does not have the intelligence to hand off calls and another part is that cell-site coverage is rather small and thus a large number of cells must be built to obtain overlapping coverage [39]. In addition, when the phone is out of range of its home base station, calls can only be made, not received. This inconvenience can be ameliorated if the individual carries a pager or uses voice mail and can thereby retrieve messages to call someone. Moreover, in Japan and France, systems are available with some incoming call capability outside the range of the home base station. However, the French system does not hand off in moving cars and the Japanese network hands off only when the car is traveling at speeds of less than thirty kilometers per hour [40].

A CT-2 handset can be used in the home or office or as a substitute for a public phone, wherever a base station has been built. As the United States Federal Communications Commission (FCC) described it: "The latter [public phone] aspect of CT-2 is referred to as 'telepoint service.'...[P]roviders set up base stations in public places, such as airports, shopping centers, restaurants, etc. Subscribers to the service access it with their personal CT-2 handsets when they are within range of a base station" [41].

CT-2 has been implemented in several countries, including the United Kingdom, Hong Kong, and Singapore. Its introduction in European countries has been less than fully successful. Reasons given for this are that customers expect a similar level of coverage to cellular and that cellular is already so inexpensive in some countries that there is no market for CT-2 [42].

3.3.4 Next-Generation Systems

The third kind of system to which PCS/PCN refers is one that is being drawn on a blank slate rather than evolving from a vehicle-based, voice transmission arrangement like cellular. However, when it is completed, it is expected to look, at least to the end user, much like the system that cellular operators are now creating by filling in the gaps in their coverage systems and expanding their range of applications. In the United States, PCS/PCN will use the 1,850- to 1,990-MHz portion of the radio spectrum [43]; in Europe, 1,710 to 1,880 MHz [44]. The FCC began auctioning blocks of PCS spectrum in December 1994. The precise technology to be used to provide services over this spectrum in the United States will be left to the market [45]. In Europe, PCS/PCN is referred to as DCS 1800. DCS 1800 licenses have been issued in the United Kingdom, Germany, and France [46]. Other than stating that the system is expected to compete with existing cellular systems, the French invitation to tender did not specify services required to be provided over this spectrum [47].

A projected standard for integration of various communications systems is the Universal Mobile Telecommunication System (UMTS) currently being developed in Europe. The goal of UMTS is to integrate the four principal types of mobile communications—cellular, wireless, paging, and private mobile radio—into one system with the pocket telephone as the principal customer equipment [48].

Much of this technology may be more advanced (and thus more expensive) than that needed to meet the needs of a developing country. However, some experts, cognizant of the way in which voice communications, data communications, and video communications (including entertainment) are converging [49] have envisioned a way that developing countries can take advantage of this convergence by using wireless communications. They propose that rural telecommunications systems can become self-supporting through establishment of community information centers composed of a computer, a printer, and a packet radio, all tied into a cellular system [50]. Local enterprises and individuals would pay for time-sharing of this equipment [51].

Even if cellular is used by a developing country only for basic voice communications rather than these more advanced services, it is helpful to understand the way in which wireless and landline systems are converging. Although some may still see wireless and landline as two separate beasts, from both a technical and business point of view, a more flexible approach will allow consideration of more options. The end result will be a more informed choice.

3.4 WHO CHOOSES?

Who should decide whether analog or digital technology is to be used in a country's system and which analog or digital air interface is appropriate? Who should determine whether the cellular system should be engineered to serve only mobile or fixed users, or both? What about which services are authorized over the frequencies allocated? In many countries, government has made these decisions; in others, they have to some degree been left to industry. The answers will depend in some part on circumstances endemic to the country involved.

As was noted, the U.S. FCC has not specified which technologies must be used over the portion of the spectrum allocated to PCS. Nor has it required existing analog cellular operations to be converted to digital or mandated the digital format that must be adopted [52]. Rather, it has left this to the industry on the theory that the market will make the right decisions.

This was not the case, however, when cellular was first introduced in the United States. The FCC, over the course of several decades, reviewed the evolving state of mobile radio technology until AMPS was fully developed and the FCC was persuaded that it would be worthwhile to allocate sufficient spectrum for cellular to become widely available [53]. Moreover, because the FCC divided the United States into hundreds of markets—some geographically quite small—which were to be served by many different operators, it wanted to ensure that a customer of one operator could use his or her phone in other systems [54].

Today, however, cellular standards are often effectively set by industry groups such as the Telecommunications Industry Association (TIA). TIA committees are composed of representatives from carriers and equipment vendors. These committees write specifications that manufacturers of cellular equipment, while not bound to do so, generally follow when designing how their equipment should be able to perform [55].

Although cellular technology is far more advanced now than when the United States first issued licenses for cellular service, there are still reasons why a government might want to specify which technology must be used by cellular licensees in its country. One reason is the need to have a technology that is compatible with those in nearby countries for the benefit of the entire region. This has been the driving force in the European Community's specification of a GSM standard [56]. Similarly, members of the Southern Africa Development Community (SADC) [57] have met to attempt to agree on a common standard [58].

Developing countries might be concerned about compatibility of their system with that of an adjoining country for another reason. It may be that the profit to be made from a cellular license to be issued in a developing country would be so small that it would fail to attract many bidders. However, if the developing country is adjacent to a larger country that is planning to build or has already constructed a cellular system, the operator of the cellular system in the larger country might be able to extend its network into the developing country at a lower cost than an entirely separate operator. It could do this by using cell sites in the adjoining country to beam into the developing country, or by switching cell sites in the developing country off switches in the adjoining country. Alternatively, an independent operator in the developing country might be able to lease switch capacity from an operator in a larger, adjacent country.

A desire to foster competition is another reason why a government might want to specify the technology to be used. For example, if a country decides to authorize two nationwide cellular systems, it might want them to utilize the same technology. This would allow a customer to switch from one system to the other—in case of dissatisfaction with service quality or pricing, for example—without incurring the cost of new terminal equipment [59].

A middle of the road option might be to ask interested bidders to propose the technology that they would like to provide and then to pick from among the submissions. This seems to have been the approach taken by Ghana, which ultimately granted six cellular licenses, although only two are operating [60]. It was also used in the issuance of the second Israeli license [61]. The disadvantage of this method is the possibility that none, or not enough of the bids would be acceptable to the government.

The government of a developing country may want to be more specific on the question of whether the cellular system(s) licensed by the country will be for mobile or fixed uses, or some combination of the two. The need for wireless to substitute for landline service in such countries would drive such specificity. Yet, to date many countries have not done so. This may be due to the perception of wireless as a "yuppie toy," [62] to a concern about alienating potential investors by making too many demands, or to a lack of knowledge about how to make such specifications [63]. Chapters 10 and 11 will discuss the ways in which a country might make such specifications.

3.5 CONCLUSION

A number of types of analog and digital technologies are available for the country contemplating the implementation of a cellular system to meet its telecommunications needs. Moreover, recent years have seen an integration of fixed and mobile applications in cellular systems that may be of particular benefit to a developing country. The choices a government makes among these options may depend in large part on the particular geographic and demographic characteristics of its country, as well as the financial resources available to it. And if the country decides to allow private interests to build and/or operate its cellular system(s)—the decision addressed in the next chapter—the choices can be limited by what private business is willing to provide.

Notes

[1] The trend in phones has been a movement from strictly mobile phones, which are installed in and cannot be removed from a vehicle, to portable phones, which can be carried in a pocket or purse. See, *e.g.,* Krister Raith, Erik Lissaker, Jan Uddenfeldt, and Jan Swerup, "Cellular for Personal Communications 2," in *Wireless Personal Communications*, Feuerstein (ed.), 1993; Calvin Sims, "A Gadget That May Soon Become the Latest Necessity," *N.Y. Times*, Jan. 28, 1990, Business Section, p. 10.

[2] For a rather technical description of the various analog and digital standards, see the ITU's "Reports of the CCIR (International Radio Consultative Committee)," 1990, p. 232, Report 742-3.

[3] Keith Bradsher, "Can Cellular Phone Companies Agree on a New Standard for Transmission?" *N.Y. Times*, Sept. 16, 1990, Business Section, p. 9.

[4] Discussion with Gerald A. V. Buttex, Telecommunications Consultant, Africa, Middle East, and Europe Regions, The World Bank, Sept. 28, 1994.

[5] But see Milo Geyelin, "Cellular Phones May Betray Client Confidences," *Wall St. J.*, Sept. 1, 1994, p. B1, col. 3 ("snoopers are already finding ways around the new [digital] technology.") Moreover, analog signals can be encrypted with special equipment.

[6] Megumi Komiya, "Fixed Wireless Local-Loop System: A New Recipe for Success?," *in Pacific Telecommunications Council Sixteenth Annual Conference Proceedings*, 1994, pp. 481, 484.

[7] For a complete list of abbreviations and the names of the technologies they represent, see Barry L. Leff, "Making Sense of Wireless Standards and System Designs," *Microwaves & RF*, Feb. 1994, p. 113.

[8] Balston & Macario, *supra* Ch. 1, note 1, at 65.

[9] *Id.* at 73–74 ("The ability to combine the 450- and 900 MHz systems using the same exchange and cell sites has been introduced by most manufacturers.")

[10] *Id.* at 114–15, 118, 137, 143.

[11] *Id.* at 31.

[12] Anthony Ramirez, "Next for the Cellular Phone," *N.Y. Times*, March 15, 1992, Business section, p. 7.

[13] Balston & Macario, *supra* Ch. 1, note 1, at 228, 334; Steven Titch, Charles F. Mason, "Digital cellular: What now?," *Telephony*, Feb. 10, 1992. (TDMA subdivides the digitized radio signal into three bit streams, or time slots, and expanded TDMA (E-TDMA) subdivides it into six.) A major personal communications services consortium in the United States has chosen CDMA as its standard. See Julie Chao, "Qualcomm's Prospects Suddenly Become Very Good," *Wall St. J.*, June 7, 1995, p. B3, col. 1. Use of CDMA technology has been mandated in South Korea. See *Telecommunications Rep'ts*, May 3, 1993, p. 48.

[14] Taken from the name of the committee of representatives from the European PTTs that was charged with selecting a pan-European standard. Balston & Macario, *supra* Ch. 1, note 1, at 154.

[15] Roger Newell, "Commentary: U.S., Foreign Battles over Standards for PCS: The Worldwide Race for TDMA, CDMA, GSM," *TR Wireless News*, Jan. 13, 1994, p. 11.

[16] "Japan, Korea and the Americas are the only GSM black spots," *FinTech Mobile Communications*, Nov. 4, 1993.

[17] Some cellular systems that are being constructed now are starting with a mix of digital and analog technology with plans to convert entirely to digital later. "BellSouth/IDB/Safra Group Gets Israel Cellular License," *Telecommunications Rep'ts*, May 16, 1994, p. 32. Digital FDMA, IS-54 TDMA, and CDMA are sometimes collectively referred to as "D-AMPS" or digital AMPS.

[18] Newell, *supra* note 15, at 13.

[19] "PCS Providers in U.S. Form North American Interest Group to Promote 'GSM' Wireless Technology," *Business Wire*, Nov. 20, 1995; "BellSouth Unit To Launch GSM Wireless Network," *Dow Jones News*, Nov. 14, 1995.

[20] *E.g.*, Nirmal Ghosh, "Globe Telecom launches $599 handphone in Manila," *Straits Times* (Singapore), Sept. 14, 1994, p. 34 (GSM subscribers may use their phones in some 50 countries under international roaming arrangement); "GSM Standard's Influence Spreads Worldwide," *Mobile Phone News*, March 1, 1993.

[21] Safeguards that prevent confidential subscriber information from being picked up by bandits with equipment that monitors the airwaves are built into the GSM network. Guy Daniels, "The European Market for Digital Cellular Communications," *Microwave Journal*, Jan. 1993, pp. 66–71. There has been some concern that the GSM encryption algorithm is so sensitive that it should not be made available to some countries. John Williamson, "GSM bids for global recognition in a

crowded cellular world," *Telephony*, April 6, 1992, pp. 37–38. A semipermanent SIM card can be plugged inside the equipment, so that a subscriber need not always remember to carry her SIM card with her. Balston & Macario, *supra* Ch. 1, note 1, at 171.

[22] Personal identification numbers are becoming more common with AMPS systems as well, as a way to deter fraud.

[23] "AT&T Corp., Cellular-Phone Coverage Expands to 35 Countries," *Wall St. J.*, Nov. 8, 1995, p. B9 (reporting on SIM cards issued to U.S. cellular subscribers for use in foreign countries); Douglas Lavin, "Deutsche Telekom, GTE Will Develop New Phone Service," *Wall St. J.*, March 8, 1995, p. B7, col. 5 (reporting on reciprocal system); "Ameritech Cellular Customers Can Use GSM Phones," *Telecommunications Rep'ts*, June 20, 1994, p. 40 (reporting on SIM cards issued to U.S. cellular subscribers).

[24] "Hungary Awards Two Digital Cellular Licenses," *Telecommunications Rep'ts*, Aug. 30, 1993.

[25] A more complete listing of countries adopting the GSM standard is contained in Appendix C, Review of Worldwide Developments, to Commission of the European Communities, Towards the Personal Communications Environment: Green Paper On a Common Approach in the Field of Mobile and Personal Communications in the European Union, COM94, 45 final, Brussels, 27.04.1994) [hereinafter, "Mobile Green Paper"].

[26] Telephone conversation with Ed Resor, Consultant, Jan. 3, 1995.

[27] See, *e.g.,* Quentin Hardy, "Motorola Wins $100 Million Contract To Provide Wireless Phones in Hungary," *Wall St. J.*, June 8, 1995, p. B12, col. 3; "Ericsson Wins $450 Million Contract for Fixed Cellular in Malaysia," *Edge*, Sept. 19, 1994.

[28] For example, a system is being installed in the Republic of Tatar that will serve primarily as a substitute for a basic wireline network, with fixed terminals in homes and businesses, and fewer than one in ten terminals dedicated to mobile use. *Telecommunications Rep'ts*, Sept. 6, 1993, p. 42.

[29] ITU, World Telecommunication Development Report 38 (1994). This ITU report was issued to participants in the first ITU World Telecommunications Development Conference in Buenos Aires in March 1994. "Vice President's Role Seen as Signal of Administration's Interest in Telecom as Catalyst for Global Development," *Telecommunications Rep'ts*, March 21, 1994, p. 12.

[30] John J. Keller, "Sprint Ready To Tap AT&T For PCS Pact," *Wall St. J.*, Dec. 8, 1995, p. A3, col. 4.

[31] Conversation with Peter Smith, *supra* Ch. 2, note 17.

[32] Komiya, *supra* note 6, at 484–85. C. Billowes, "Some Background to and Experiences in Rural Telecommunications in the Third World," in *Second International Conference on Rural Telecommunications*, London: Institution of Electrical Engineers, 1990.

[33] Rudi Westerveld, "Cost Effective Rural Communications Using Fixed Cellular Radio Access," in B. A. Kiplagat & M.C.M. Werner, *Telecommunications and Development in Africa*, 1994, p. 199, 205. The Westerveld article describes how a mixed use—mobile and fixed—system can be engineered to serve rural areas.

[34] In the United States, the Federal Communications Commission (FCC), the agency that regulates cellular service, for years mandated that, with the exception of basic exchange telephone radio service ("BETRS"), fixed point-to-point services over cellular frequencies could only be offered on an "incidental" basis. Report and Order, 65 RR 2d 985, 995–6 (1988). BETRS is designed to serve as an extension of basic exchange service in areas where inadequate or no basic is offered by a landline carrier. For current policy on provision of fixed services over cellular frequencies, see 47 C.F.R. §§22.901, 24.5.

[35] Some do not distinguish between microcells and picocells, but simply refer to any cell with a coverage radius of less than one kilometer as a microcell. "Cellular for Personal Communications," *supra* note 1. Others refer to a picocell as one with a coverage radius of, for example, 30–250 meters. Hans van der Hoek, "The New DECT Standard for Cordless Communications," *Telecommunications*, April 1993, p. 37.

[36] Notice of Propose of Rulemaking, in the Matter of Amendment of the Commissions Rules to Permit Flexible Service Offerings in the Commercial Mobile Radio Services, FCC 96-17 (Released January 25, 1996).

[37] For years, U.S. cellular frequencies could be used for "auxiliary" services such as data (voice is considered "conventional") transmissions only on a secondary basis. Report and Order, *supra* note 34, at ¶ 2. For current policy, see 47 C.F.R. §§22.901, 24.5.

[38] Laurie Flynn, "The Executive Computer: 3 Ways to be Unplugged Right Now," *N.Y. Times*, Dec. 4, 1994, § 3, p. 11; "BAM to Deliver Cellular Digital Packet Data," *Bell Atlantic World*, Third Quarter 1993, p. 1. See Bruce Caldwell, "GTE Prepares CDPD Rollout—Announces successful interoperability test," *Informationweek*, Nov. 28, 1994; "Why Your Company Should Think Wireless," *Investor's Business Daily*, Nov. 16, 1994, p. A4; Jeffery Schwartz, "US West to Offer Wireless Data Services Via Voice Nets, Not CDPD," *Communicationsweek*, Oct. 31, 1994. As the *New York Times* article cited above describes, CDPD is not the only packet data technology available.

[39] John Williamson, "The wireless industry erupts worldwide," *Telephony*, Aug. 3, 1992, p. 26.

[40] Jenny Walker, "Tiger economies beat forecasts," *Fin'l Times*, Nov. 27, 1995, Special Section on Mobile Communications, p. 4., col. 1; David P. Hamilton, "Japan Rolls Out Entries in PCS Market Meant to Rival Bigger Cellular Phones," *Wall St. J.*, July 3, 1995, p. A3A, col. 2 (discussing Japanese "Personal Handy Phone"); "Romancing the Bi-Bop: the French and Their Phone," *N.Y. Times*, April 23, 1995, § 1, p. 49 (calls can be forwarded to the portable phone if the caller will be in one place for a long time); Michael Paetsch, *Mobile Communications in the U.S. and Europe: Regulation, Technology, and Markets*, Norwood, MA: Artech House, 1993, p. 327. (The French Bi-Bop system allows user to key in a code number on his or her handset, and incoming calls will be routed to the specific base station that received the code.)

[41] Notice of Inquiry, In the Matter of Amendment of the Commission's Rules to Establish New Personal Communications Services, at p. 3, n. 3, GEN Docket No. 90-314 (1990).

[42] MalarkyTaylor Associates/Economic and Management Consultants, Inc., Asian Cellular Markets, Vol. 1, 1993, p. 17; "Finnish closure leaves only two telepoint systems in Europe," *FinTech Mobile Communications*, June 16, 1994.

[43] Memorandum Opinion and Order, In the Matter of Amendment of the Commission's Rules to Establish New Personal Communications Services, 9 FCC Rcd. 4957 at ¶ 26 (1994).

[44] Daniels, *supra* note 21, at 76; Balston & Macario, *supra* Ch. 1, note 1, at 208.

[45] Memorandum Opinion and Order, *supra* note 43, at ¶ 3.

[46] Thomas Kamm, "Boygues Wins a Second Contract, Posts Profit Rise and Will Seek New Capital," *Wall St. J.*, Oct. 6, 1994, p. A14, col. 1; Balston & Macario, *supra* Ch. 1, note 1, at 207; Kavita Bowry, "A Review of European Communications," *Microwave Journal*, Oct. 1993, p. 64.

[47] Invitation to Tender, for setting up a public radiotelephone network, aiming to develop a personal communications service on French territory (Feb. 1994).

[48] Cengiz Evci and Vinod Kumar, "Pan-European Project for Third Generation Wireless Communications," in Jack M. Holtzman & David J. Goodman, *Wireless Communications: Future Directions*, 1993.

[49] *E.g.*, "Geoworks to Provide Ericsson with System for 'Smart Phones,'" *Wall St. J.*, Jan. 23, 1996, p. B7 (describing cellular phones that supplement voice communications with features like faxing, Internet connections, and electronic mail); Diane Mermigas, "Virtual Reality (Part 1 of 2): Moving Beyond the Dreams of the Communications Revolution," *Electronic Media*, Nov. 7, 1994, p. 38; "Multimedia: Sprint Unveils Multimedia Test Track in Silicon Valley," *Edge*, Oct. 224, 1994.

[50] For early uses of packet radio in developing countries, see Willem Zijp, "Improving the Transfer and Use of Agricultural Information," *World Bank Discussion Paper 247*, at 76–80 (1994).

[51] Telephone conversation with Ron Epstein, Volunteers in Technical Assistance, Nov. 23, 1994. Mr. Epstein suggests that the equipment for such an arrangement would cost about $1,500–$1,600 and that such a center could be self-sufficient if about twenty to thirty thousand people live within one-half hour of it. Countries that he indicates have attempted or are attempting to establish such centers are Uganda, Turkey, Madagascar, and India. Mr. Epstein is working with the World Bank to promulgate such centers in developing countries. He also indicated that the European Union is examining a proposal for such centers, referred to as Community Educational Utilities. Telephone conversation with Ron Epstein, Feb. 17, 1995. See Saunders, *supra* Ch. 2, note 1, at 354–55 (discussing similar concept of "telecottages" in Scandinavia).

[52] Balston & Macario, *supra* Ch. 1, note 1, at 226.

[53] Philip Palmer McGuigan, David M. Connors, and Kenneth L. Cannon II, "Cellular Mobile Radio Telecommunications: Regulating an Emerging Industry," 1983 *B.Y.U.L. Rev.* 305, 307–10, 1983; Report and Order, In the Matter of An Inquiry Into the Use of the Bands 825-845 MHz and 870-890 MHz for Cellular Communications Systems; and Amendment of Parts 2 and 22 of the Commission's Rules Relative to Cellular Communications Systems, 86 F.C.C.2d 469, 49 R.R.2d 809 (1981). It should be noted that certain aspects of the McGuigan article (such as the statement that the FCC would not use lotteries to award licenses) are out of date and contradicted by subsequent events.

[54] McGuigan, *supra* note 53, at 318.

[55] Comments of Jim Akerhielm, Director, Network Planning, Bell Atlantic Mobile Systems, Inc., Aug. 31, 1994.

[56] Directorate-General, Telecommunications, Information Market and Exploitation of Research, Overview of the Green Paper on Mobile and Personal Communications and Extract of the Positions Proposed, §§ II and III [hereinafter "Overview"].

[57] The SADC comprises Angola, Namibia, South Africa, Botswana, Zambia, Lesotho, Swaziland, Zimbabwe, Malawi, and Mozambique. Bill Keller, "Southern Africa's Old Front Line Ponders Its Future in Mainstream," *N.Y. Times*, Nov. 20, 1994, p. A1, col. 4.

[58] Although it was reported that the SADC decided to adopt AMPS because it was less expensive than GSM, with the hope of migrating to GSM, South Africa subsequently issued two GSM (rather than AMPS) licenses. "The Cellular Switch is On In South Africa," *Africa Communications*, July/Aug. 1994, p. 22; "Africa Works Toward a Common Cellular Standard," *Mobile Phone News*, March 8, 1993; *Mobile Phone News*, April 19, 1993. The issuance of the GSM licenses immediately may have occurred because GSM cannot be overlaid on AMPS.

[59] Balston & Macario, *supra* Ch. 1, note 1, at 114.

[60] Conversation with Jim Cowie, Senior Telecommunications Specialist, Industry and Energy Division, The World Bank, Sept. 28, 1994. Ghana also allowed bidders for the licenses to specify what coverage area they wished to serve. *Id.*

[61] Alan Stewart and Alan Pearce, "PCS: first, find your market," *Communications Int'l*, Oct. 1994, p. 78.

[62] Telephone conversation with Ed Resor, Consultant, Oct. 3, 1994.

[63] Conversation with Jim Cowie, *supra* note 60.

CHAPTER 4

▼▼▼

IS THE CELLULAR OPERATOR TO BE A GOVERNMENT AGENCY OR INVESTOR?

4.1 INTRODUCTION

The degree of private industry involvement in the implementation of cellular service varies widely from country to country. A private company may simply supply the equipment to the government for it to construct its own system. Or it may build a turnkey system that it hands over to the government for operation. This is often referred to as a *build and transfer* (BT) project. The BT project may involve a number of years of private operation, either before or after the transfer. Schemes in which the operation occurs before the transfer are often referred to as *build, operate, and transfer* (BOT) or *build, own, operate, and transfer* (BOOT). Those in which the operation occurs after the transfer may be called *build, transfer, and operate* (BTO) projects [1]. Or the private company may obtain a license—possibly as part of a consortium in which a government agency may also be a party—to construct, own, and operate the cellular network without any obligation to turn the system over to the government. Which alternative is more likely to bring the necessary services to the public at a reasonable cost? Are national interests threatened by having essential telecommunications services in the hands of private investors, particularly foreign investors? Examples from which we might begin to learn the answers to these questions are discussed below.

The form of a telecommunications operator can fall anywhere along the public-private spectrum. It may be part of a government agency that also has responsibility for regulating telecommunications, or it may be organizationally separate. As a separate wholly owned state entity, it may be formed under a statute created specifically for it or under the country's general company law. Or it may be only partly owned by the state, which may be either a majority or a minority equity holder. And finally, there is full-scale privatization, without any governmental participation at all [2].

During the 1980s, industrialized and developing nations both participated in a wave of telecommunications privatizations [3]. Originally, telecommunications was viewed as a natural monopoly and "a typical public service" due to its economies of scale, political and military sensitivities, and large externalities [4]. Two primary trends, however, changed this view in industrialized countries: increased demand and technological innovation. The cost of transmitting and processing information decreased, which altered the cost structures of telecommunications and other industries, created new ways of meeting telecommunications needs at lower cost, reduced dependence of users on established telecommunications operating companies, and increasingly integrated information and telecommunications technologies and services. Moreover, the political climate became conducive to privatization [5]. These factors also came to bear in developing countries [6]:

- State monopolies had reached the limit of their ability to accelerate the supply of telecommunications services. In particular, governments realized that they could not provide the huge amounts of capital required to catch up with demand.
- In recent years, many developing countries have begun to adopt market-oriented economic strategies, including measures to liberalize trade, promote competition, deregulate financial and capital markets, reduce restrictions on foreign investment, and restructure private enterprises. In order for these broad economic reforms to be effectively implemented, adequate telecommunications infrastructures urgently needed to be developed...
- Popularly elected governments found that public dissatisfaction with service and, in many countries, extensive corruption of telephone company personnel resulted in widespread support for major reform initiatives.
- Parallel changes in the telecommunications sectors of industrial countries raised international awareness of a wide range of sectoral policy issues and options and demonstrated the viability and increased political desirability of alternatives to state monopoly.
- Telecommunications operating companies in industrial countries, repositioning themselves in their own changing domestic and regional mar-

kets, aggressively pursued new business opportunities in developing countries. Also, foreign banks sought to shift their exposure in highly indebted developing countries from nonperforming loans to new investment opportunities.

It is likely that the privatization of telecommunications in developing countries has also resulted to some degree from conditions attached to economic assistance offered by lenders. For example, the IBRD (World Bank) has made clear its policy that infrastructure should be run like a business and that greater private investment is the preferred way to sharpen efficiency [7].

As the wave of privatizations continues, so does the private versus public debate [8]. A shift away from government involvement can provide a greater commercial orientation for the enterprise that encourages greater economic efficiency. However, this opportunity may not be realized if the operator takes advantage of monopoly powers to charge high prices without improving service or efficiency. Moreover, "national goals of social, economic, and political integration may be ignored or undermined" [9]. Appropriate regulation may be able to mitigate these problems, although inappropriate regulation can exacerbate them. Building regulatory institutions in countries without a strong regulatory tradition can be a slow and arduous task [10].

Government-owned telecommunications entities in developing countries have been limited in their ability to meet the demand for their services for several reasons [11]:

- First, sector policies are often inadequate. The enterprises are usually viewed as traditional public sector utilities without regard to their business character and resource mobilization potential. In particular, they often lack financial and administrative autonomy, have little incentive to improve performance, are not allowed to remunerate and promote staff as necessary to attract and retain specialized personnel, are denied tariffs that reflect costs, cannot access capital markets despite being profitable businesses, and suffer from government interference in management.
- Second, telecommunications investment is constrained by the countries' limited capital resources, especially in foreign currency...Furthermore, like other public or parapublic entities, telecommunications enterprises are subject to investment ceilings related to broader efforts to contain public sector spending.
- Third, weaknesses in the organization and management of telecommunications enterprises result in high expansion and operating costs, poor maintenance, and limited capability for project preparation and implementation.

Cellular in particular has often been privately provided, frequently from its inception. It is distinguishable, both in fact and perception, from landline service. Cellular, in contrast to landline, is often perceived as a luxury that need not be under government control. Although this perception fails to appreciate the potential of cellular, it has resulted in governmental willingness to relinquish control over it in some cases. Moreover, cellular has more easily been made competitive than local landline service historically could be [12].

How does one measure whether a telecommunications operator, public or private, is successful? One can examine its financial performance—how profitable it is—but this does not tell whether the demand for services is being met. On the other hand, if demand must be met in a way that the operator cannot be a commercially viable enterprise, one cannot expect to attract profit-seeking investors and obtain the concomitant benefits. Profitability figures are also very difficult to obtain.

Penetration rates—what percentage of the population has subscribed to service—provide a rough estimation of the success of a telecommunications provider [13]. However, penetration rates should be utilized with the caveat that they do not disclose the distribution of service among income groups. Knowing, for example, that an operator has a 25% penetration rate does not disclose whether those with phone service are the 25% wealthiest people in the country or whether subscribership is evenly distributed among income groups.

4.2 EGYPT

One country that has decided to maintain a high degree of control over its cellular systems, allowing private industry only to build the network and then hand it over to the government for operation, is Egypt. ARENTO, the Arab Republic of Egypt National Telecommunications Organization, is the state-owned entity that operates the country's existing analog system [14]. The telecommunications industry, including cellular, is the exclusive domain of ARENTO [15].

As of early 1993, according to the vice chairman of ARENTO, the system provided service to 7,500 subscribers and covered the greater Cairo, Delta, and Suez Canal areas and parts of the north coast of Egypt [16]. Figures from the United States Department of Commerce are lower, indicating approximately 5,600 total cellular subscribers in Egypt as of December 31, 1993 [17]. Using the ARENTO numbers and a population figure of 59.5 million [18] results in a cellular penetration rate of slightly more than one hundredth of one percent (.0126%) [19]. This means that for every 100,000 people, only about 13 are subscribers to cellular service. This low penetration rate may indicate that even the more well-regarded government agencies [20] will find it difficult to operate a telecommunications system as successfully as a private business can. In addition, this low penetration rate may in part be the result of other factors, such as income levels. Egypt is considered a low-income country by the ITU [21].

In 1992, ARENTO invited several international telecommunications companies and consortia to bid for a cellular license to build and operate a privately held GSM system [22]. The operator would have worked under the general supervision of

ARENTO and paid an annual fee to ARENTO [23]. Thirteen bidders were short-listed [24], and the Egyptian government announced that it would choose from among seven final bids [25]. However, in April 1994, ARENTO Chair Mahmoud Al-Soury canceled the 1992 tender. Two reasons were given for this action. First, the tender process had been plagued with "rivalries and rumor mongering" among local supporters of the seven original bidding consortia that had caused "pressure and embarrassment" to ARENTO [26]. Second, ARENTO decided that it could earn large profits by operating the system itself [27]. In its place, the Egyptian government is issuing invitations to tender bids for a turnkey system in phases, each of which will cover a different area of the country [28].

4.3 COLOMBIA

Colombia presents an interesting example of the trend toward privatization in Latin America. The Colombian government divided the country into three regions and authorized two providers—an A band and a B band operator—in each region. The B band licenses were awarded to "private" companies and the A band authorizations to "mixed" companies. A mixed company is one that includes public (government) participation [29]. Bidding was held for both the A and B licenses, and the A band winner was required to pay at least 95% of the amount of the winning bid for the B band authorization [30].

The decision by the Colombian government to allow private participation in cellular was not uncontroverted. In September 1992, the Colombian Council of State had ruled that a law authorizing foreign investment in the telecommunications sector was unconstitutional [31]. Subsequently, Law 37 of 1993 was passed to provide the framework for the provision of cellular in Colombia [32]. Law 37 authorizes foreign participation in cellular operators, without a limit on percentage ownership, through joint venture contracts, but stipulates that the foreign companies must be established in Colombia and must be experts in the provision of mobile cellular telephony. Licensees are prohibited from transferring equity or ceding the license for three years. The law also requires the companies that are awarded cellular licenses to become limited liability public corporations within five years, which means that they must be registered with the Colombian Stock Exchange. Public corporation law places a 30% limit on capital concentration by any individual or stockholder company. Law 37 also requires winners of cellular licenses to sell 10% of their capital to "social benefit organizations," such as employees, foundations, and nonprofit organizations, within four years of the grant of concession. The social benefit organizations will have three years to pay for the shares and are excluded from having to place their shares on the stock exchange [33].

After approximately one year of service, cellular penetration in Colombia was estimated at 0.45%. The relatively high penetration rate is attributed to several factors: poor condition of the existing phone system, awareness of cellular due to its existence in other countries, competition in pricing, security against violence, and a booming economy [34]. Colombia is considered by the ITU to be a lower middle income country. Unfortunately, Colombia's drug trade likely also has contributed to the growth of cellular [35].

4.4 CZECH REPUBLIC AND SLOVAKIA

Many countries have limited foreign participation in cellular ventures to less than 50%. An example of this is concessions that were granted by Czechoslovakia: now the Czech Republic and Slovakia.

In 1990, the Federal Ministry of Posts and Telecommunications of what was then the Czech and Slovak Federative Republic (the "Ministry") granted licenses to provide cellular service in the 450-MHz range and packet data service in the Czech and Slovak Republics. The licenses were granted to two joint ventures owned 51% by the Post and Telecommunications Administration (PT) in each republic and 49% by Atlantic West B.V. (AWBV), a Netherlands corporation equally owned by Bell Atlantic International, Inc. and U S West International, Inc. [36]. The two joint ventures provide service under the names Eurotel Prague and Eurotel Bratislava.

Although the U.S.-controlled party to each joint venture owns less than 50% of the equity interest in the venture company, it has equal representation on the companies' boards of directors. AWBV and the PT each appoint two directors to the board of four directors. Appointment of the chair of the board, who is one of the four directors, alternates yearly between the two parties, with the PT having the right to appoint the chair for the first year [37].

AWBV was given the exclusive right to appoint the managing director, who is the principal executive and operating officer of the venture, for the first six years [38]. After this period, appointment of the managing director is subject to approval of the board. Appointment of the other officers is divided between the parties. The PT has the exclusive right to nominate the first deputy managing directors (one for the cellular system and one for the data network), marketing director, and personnel director. AWBV has the exclusive right to nominate the first operations director and finance director. The managing director nominates all other officers. All of these nominations are subject to board approval [39]. As local employees were trained over the first four years of the ventures, the participation of U.S. employees has declined [40].

According to figures from the U.S. Department of Commerce, at the end of 1993 Eurotel Prague had 12,300 subscribers and Eurotel Bratislava had 3,090. Population figures for the Czech Republic and Slovakia were 10,389,256 and 5,375,501, respectively. Both republics are designated as lower middle income by the ITU, although the Czech Republic has recently done better economically than Slovakia [41]. Thus, cellular penetration rates were approximately one tenth of one percent (.118%) for the Czech Republic and one twentieth of one percent (.057%) for Slovakia.

4.5 MEXICO

Another country that has long limited foreign participation in cellular is Mexico [42]. Originally, Mexico permitted no foreign investment in the mobile telecommunications sector. However, restrictions on foreign investment were relaxed in 1990 to allow up to 49% foreign ownership [43]. Then, in 1995, Mexican law was amended to authorize the National Commission on Foreign Investment to permit ownership of cellular licenses in excess of 49% [44].

The government has divided the country into nine regions, with two cellular providers in each region. One license in each region is owned by Radiomóvil DIPSA, S.A. de C.V. (Telcel), the cellular affiliate of Teléfonos de México, S.A. de C.V. (Telmex), the recently privatized telephone company that provides, among other things, landline telephone service in Mexico [45]. As in the United States, the license held by the landline company's affiliate is referred to as the "wireline" system. Most of the nonwireline systems in Mexico are held in whole or part by other major companies. Grupo Iusacell, S.A. de C.V. (Iusacell), the largest nonwireline cellular carrier in Mexico [46],currently holds the concessions for Region 9—which includes Mexico City—as well as Regions 5 (Guadalajara), 6 (central Mexico), and 7 (southeast Mexico) [47]. Nonwireline licenses in the regions along the United States-Mexico border (Regions 1, 2, 3, and 4) are owned by alliances in which the U.S. company Motorola, among others, has heavily invested [48]. Region 8 covers the southernmost and easternmost parts of Mexico, including the troubled area of Chiapas.

Iusacell itself is owned in large part by Bell Atlantic Corporation. Bell Atlantic holds a 42% equity stake, with a 44% voting interest [49]. However, certain significant matters can only be decided by a supermajority vote, providing the minority U.S. shareholder with some protection [50]. Bell Atlantic also obtained the right to appoint Iusacell's chief financial officer, the two officers responsible for strategic planning and network planning, and the secretary of the company.

Cellular penetration rates in Mexico vary depending on the region. Bell Atlantic reported that as of March 31, 1994, shortly before it acquired its stake in Iusacell, the company's penetration rate in the region that included Mexico City was 0.42%; the penetration rate in the remaining Iusacell regions was 0.13% [51]. These figures are for the nonwireline side only, rather than for cellular services entirely, in contrast to the numbers given above for Egypt and the Czech and Slovak Republics, where there is only one carrier. Nationally, combined cellular penetration in Mexico as of June 1993, after four years of operations, was approximately 0.44%. Since 1993, the number of customers has increased at a substantial rate of growth [52]. Iusacell announced that the number of its customers increased from 157,151 to 202,157 for the year ending June 30, 1995, a 29% growth rate [53]. Mexico is considered by the ITU to be an upper middle income country.

4.6   UNITED STATES

The United States also places restrictions on what it less than tactfully refers to as "alien" ownership. Section 310(b) of the Communications Act of 1934, as amended, provides that no common carrier radio license [54]:

> shall be granted to or held by—
>
> (1)   any alien or the representative of any alien;
>
> (2)   any corporation organized under the laws of any foreign government;
>
> (3)   any corporation of which more than one-fifth of the capital stock is owned of record or voted by aliens or their representatives or by a foreign government or representative thereof or by any corporation organized under the laws of a foreign country.
>
> (4)   any corporation directly or indirectly controlled by any other corporation of which more than one fourth of the capital stock is owned of record or voted by aliens, their representatives, or by a foreign government or representative thereof, or by any corporation organized under the laws of a foreign country, if the commission finds that the public interest will be served by the refusal or revocation of such license.

This limitation was originally enacted because of concerns related to radio broadcast licenses. The fear was that alien control would allow U.S. citizens to be bombarded with foreign propaganda [55]. Nevertheless, the statute reached radio common carriers as well, and cellular operators are considered common carriers. Section 310(b) is specifically made applicable to cellular authorizations by the FCC's regulations [56]. The FCC has revoked a cellular license in a case in which the alien ownership prohibition was violated [57].

As Section 310(b)(4) states, the FCC can approve greater than 25% foreign involvement up the corporate chain from the actual licensee where it is in the public interest. In order to determine whether such public interest exists, the FCC will examine whether effective competitive opportunities exist in the home cellular market of the foreign carrier seeking to acquire the interest. If U.S. companies are allowed by the law of a foreign country to acquire a controlling interest in cellular carriers in that foreign market, then companies whose home market is in that foreign country will be allowed to exceed the 25% benchmark in their ownership of U.S. cellular interests, absent significant *de facto* barriers to competition. If U.S. companies are not permitted to acquire controlling interests in the foreign country, then the FCC will allow the 25% benchmark to be exceeded only to the same degree that a U.S. company can acquire more than 25% of a cellular carrier in the foreign market. (For example, if a particular country allows U.S. companies to acquire a noncontrolling 40% interest in a cellular licensee, then an investor from that country would be able to acquire up to a noncontrolling 40% interest in the holding company of a cellular licensee in the U.S. market.) The kinds of *de facto* barriers that the FCC will consider even if *de jure* access is available are (1) whether there exist reasonable and nondiscriminatory charges, terms, and conditions for interconnection to the foreign network, as well as means to monitor and enforce these provisions; (2) whether there are safeguards against anticompetitive practices, such as the existence of cost-allocation rules to prevent cross-subsidization, timely and nondiscriminatory disclosure of technical information needed to use or interconnect with carriers' facilities, and protection of carrier and customer proprietary information; and (3) whether there is an effective regulatory framework in the foreign country, including separation between the foreign regulator and operators and fair and transparent regulatory procedures. The FCC will also consider other public interest factors: "the general significance of the proposed entry to the promotion of competition in the U.S. telecommunications market; any national security, law enforcement, foreign policy, and trade concerns raised by the Executive Branch; and the extent of alien participation in the applicant's parent corporation..." The FCC has stated that among the goals that it hopes to achieve with this policy is to encourage foreign governments to open their telecommunications markets to U.S. companies [58].

In its first application of this policy, the FCC in 1996 approved a foreign investment of up to 28%, including investments of 10% each by Deutsche Telekom and France Telecom, in Sprint Corporation. Although the FCC did not find that Germany and France offer effective competitive opportunities, the FCC took note of plans by the governments of these countries to liberalize their telecommunications markets in the next few years. The FCC also noted that allowing the investment would provide Sprint with the funds to be a more effective competitor in the United States, which would benefit U.S. consumers of telecommunications services. Nevertheless, the FCC made its approval conditional and stated that it would take further action within two years if the liberalization in Germany and France did not occur as planned [59].

Cellular penetration in the United States averaged about 7% in late 1994. By mid-1995 it was around 10% [60]. Cellular service in the United States has always been privately provided (that is, not provided by government entities).

4.7 RESTRICTIONS ON FOREIGN INVESTMENT

As the above examples illustrate, restrictions on foreign investment in cellular are common. While the idea of foreign investment often draws a negative visceral reaction from citizens of the country in which the investment is made, it is important to examine exactly what those in favor of restrictions on such investment actually have to fear.

Developing countries in particular often have generalized concerns about foreign investment. This is perhaps not surprising given the history of colonialism in such countries, under which industrialized countries used them as sources for raw materials and markets for their products [61]. But there are fundamental differences between colonialism (or neocolonialism), which furthers national interests, and foreign private investment [62]. The commercial entity has profit maximizing goals that are not the same as the imperial goals sometimes held by governments. And while colonialism brought about the depletion of irreplaceable natural resources (e.g., oil), the provision of telecommunications services does not carry the same danger [63].

It has been noted, however, that when a commercial enterprise interested in doing business in another country enlists the persuasive powers of its home government on its behalf, the perception of an identity of interests may be increased [64]. Many countries do promote the interests of their domestic businesses abroad, because this will strengthen the home country's economy and create more jobs [65]. However, if the process is founded on the premise that the developing country's choice is to be uncoerced and informed, it can not be equated with colonialism or neocolonialism.

This is not to say that developing countries should not be aware of the costs, as well as the benefits, of foreign investment. Profits of telecommunications ventures are likely to be repatriated to the investor's home nation, rather than kept in the country where they might be reinvested in the local economy [66]. Or a country's economy may become so dependent on foreign cash that a rapid outflow due to severe reaction to developments in that country may cause tremendous disruption [67]. Currency and investment restrictions may prevent this, but may also discourage desired foreign investment. A sophisticated foreign investor with superior knowledge and resources may also attempt to "take advantage of" the government of a developing country in negotiations over the terms of the investment. To some extent this can be avoided if the government can retain equally sophisticated lawyers and investment advisers [68]. This is obviously easier for a large country to accomplish than a smaller one.

On the other hand, insisting on domestic equity ownership may impose additional costs on the government. A foreign investor's costs may be increased if it must obtain domestic partners. Particularly in developing countries, domestic companies do not always have sufficient capital to satisfy the contribution they must make in exchange for their equity interest in the operating entity. They may depend on financial assistance from the foreign investor in this situation. Money that the foreign investor must apply toward support of its domestic partner is money that cannot be paid to the government for the cellular license or that will be passed on to customers. And inability to find domestic partners that are qualified to participate in the venture may simply discourage bidders entirely.

Even if fears of foreign investment generally are at least partially attributable to emotional rather than rational reactions, are they more justified in the context of telecommunications, specifically cellular, because more is at risk? (This issue is also relevant with respect to satellites, over which developing countries have less control.) Telecommunications is considered a strategic sector of the economy [69] for all the reasons discussed at the outset of this book. Is it dangerous to have what may plausibly be referred to as a country's lifeline in the hands of foreign private investors rather than local private investors?

Before answering this question, it is important to separate the question of which private investor—domestic or foreign—should be allowed to have control, from the question of whether control should be in the hands of government rather than a private investor. Sometimes, what seem to be arguments against foreign ownership are really, at least in part, arguments against private ownership generally. For example, nondemocratic regimes that fear mass access to communications will loosen their hold on the populace would be as threatened by local private ownership as by foreign private ownership, except to the extent that local owners may be more easily intimidated by the government.

Foreign investors may be less subject to internal political pressures than domestic companies and perhaps less concerned than a local investor about the well-being of the local populace; conversely, they may be more subject to pressure from their own government and more interested in the success of their own country's economy. These influences could affect their operation of another country's cellular system. For example, the operator might be inclined to award contracts for network equipment (cell sites and switches) or telephones (the end user's equipment) to one of its affiliates or to another company based in the operator's home country, at an inflated price or at lower quality than might be obtained from another vendor.

There are several ways for developing countries to minimize the possibility of being held hostage to the operator's own agenda. One is to authorize competitors to provide the same service. Another is through regulation. A third is to make it a condition of the license that the foreign investor train local managers to assume responsibility for the business after a certain period of time. Other ways of reducing the influence of the foreign investor are to require it to reduce its interest by a public stock offering within a specified number of years or to require that a majority of the board of directors of the telecom corporation be local business persons [70].

4.8 CONCLUSION

Private investors clearly have valuable contributions to make to cellular ventures. In countries with insufficient capital and domestic expertise, foreign private investors are particularly desirable [71]. Some governments have *required* foreign participation in consortia seeking cellular licenses [72]. Privatization has produced important benefits in telecommunications, including cellular. However, governments often choose to retain a measure of control over telecommunications, including cellular operations, out of concern that private investors, particularly foreign ones, do not completely share their agenda. In doing so, they must consider whether the potential loss in economic efficiency can be justified by the possible advancement of social good. Investors must be alert to these concerns and prepared to make at least some concessions that will assuage them.

Notes

[1]   Smith and Staple, *supra* Ch. 1, note 7, at 54–56.
[2]   Timothy E. Nulty, "Emerging Issues in World Telecommunications," *in Restructuring and Managing the Telecommunications Sector: A World Bank Symposium,* Björn Wellenius, Peter A. Stern, Timothy E. Nulty, and Richard D. Stern, eds., 1989, p. 16.
[3]   Saunders, *supra* Ch. 2, note 1, at 303–31.
[4]   *Id.* at 305. An externality occurs when "the price system...fail[s] to register all the benefits and costs associated with the production and consumption of certain goods and services. That is, some benefits and costs are external to the market in that they accrue to parties other than the immediate buyer and seller." Campbell R. McConnell, *Economics* 97, Seventh ed., 1978. With respect to telecommunications, "[o]ne potential implication of the presence of significant externalities is that competitively determined prices may be too high and therefore may not lead to efficient-sized networks or to the socially 'optimal' volume of calling." Smith and Staple, *supra* Ch. 1, note 7, at 15.
[5]   Saunders, *supra* Ch. 2, note 1, at 304–05, 322.
[6]   *Id.* at 309–10.
[7]   *E.g.,* "Investing in development," *The Economist*, June 25, 1994.
[8]   *E.g.,* Douglas Lavin, "France Puts Off Telecom Privatization," *Wall St. J.,* July 12, 1995, p. A10, col. 1; Matt Moffett, "Slowly, Brazil Sets Stage to Free Economy," *Wall St. J.,* May 19, 1995, p. A10, col. 1.
[9]   Saunders, *supra* Ch. 2, note 1, at 309-10.
[10] *Id.* at 323.
[11] Björn Wellenius, "Beginnings of Sector Reform in the Developing World," in *Restructuring and Managing the Telecommunications Sector: A World Bank Symposium, supra* note 2, at 91. See also Saunders, *supra* Ch. 2, note 1, at 64–81; Björn Wellenius and others, "Telecommunications: World Bank Experience and Strategy," World Bank Discussion Paper 192, 1993.
[12] Competition in landline service is a technological, economic, and political issue. The local landline network has historically been a monopoly bottleneck because it did not make economic sense to run more than one wire to each person's telephone and it seemed efficient for one entity to own the entire network. Charles H. Kennedy, *An Introduction to U.S. Telecommunications Law*, Norwood, MA: Artech House, 1994, p. xvi. Although technology is changing with respect to competition in local landline service, established monopoly providers of local service have resisted competition, particularly where they are not free to enter their competitors' markets (such as cable television). And in some countries, the provider of local phone service is an arm of the government, which can make political change difficult. The European Union has decided that, with a few exceptions, state landline telephone companies in member states must lose their monopolies over wires and switches as of Jan. 1998. "World Wire," *Wall St. J.,* June 14, 1995, p. A15; "EU to Free Telecom Networks," *Wall St. J.,* Nov. 18, 1995, p. A14. (In contrast, it is considering issuing an order for member states to open mobile telephony to full competition by January 1, 1996. *World Wire,* June 20, 1995, p. A9.) One reason why it is difficult to open these historical landline monopolies to competition at this particular time is that governments are trying to privatize them.

They believe it will be difficult to find enough capital to invest in these companies if they are going to be subject to competition immediately.

[13] *E.g.,* "SBC's Profit Climbs 23.7%," *N.Y. Times,* Oct. 18, 1994 (unit was first in U.S. to achieve a 7% penetration rate); "ALLTEL Earnings Per Share Up 18 Percent," *PR Newswire,* July 20, 1994 (cellular penetration is a "key operating result").

[14] "C&W and Vodafone in the running for GSM cellular license in Egypt," *FinTech Mobile Communications,* Nov. 5, 1992; Paul Spiers, "Cellular is big business in Egypt," *Telephony,* Feb. 3, 1992, p. 30.

[15] Samir H. Hamza and Ruairidh W. Campbell, "Telecommunications Law in Egypt," in *Telecommunications Law in Europe,* Joachim Scherer (ed.), 1993, p.45–49.

[16] Mohamed Selim, Vice Chairman, ARENTO, "Egypt: International Telecommunications," *Africa Communications,* March/April 1994, p. 36.

[17] However, the ITU gives a figure of 7,550 for 1993.

[18] "Millicom—AT&T to Support Egyptian Cellular Deal," *Extel Examiner,* Oct. 7, 1993.

[19] The United States Department of Commerce figures represent a penetration rate of .0094%.

[20] See., *e.g.,* Vineeta Shetty, "The famished road to a full network; telecommunications systems in Africa," *Communications Int'l,* April 1994, p. 52.

[21] However, Sri Lanka, which is also a low income country, has a noticeably higher cellular penetration rate: .13%. Ch. 5, *infra.*

[22] "Millicom, Inc. announces an agreement between MIC and AT&T for support in proposed cellular license in Egypt," *Business Wire,* Oct. 7, 1993.

[23] U.S. Department of Commerce Incoming Telegram, April 1994, Subject: Egypt Changes Plans for Cellular System. Moreover, the U.S. Trade and Development Agency offered the Egyptian Minister of Transport, Communications, and Civil Aviation a grant of up to $263,000 if one of the U.S. bidders were selected. U.S. Department of Commerce Incoming Telegram, March 1994, Subject: TDA Grant Proposal for Cellular Phone Project.

[24] "Egypt goes for GSM," *FinTech Mobile Communications,* Oct. 8, 1992.

[25] "International Phone Update, 10/11/93," *Newsbytes News Network,* Oct. 11, 1993.

[26] "Egypt Halts Tender for Private Cellular System," *Global Telecom Rep't,* May 2, 1994; U.S. Department of Commerce Incoming Telegram, April 1994, *supra* note 23.

[27] "Egypt Halts Tender for Private Cellular System," *Global Telecom Rep't,* May 2, 1994; U.S. Department of Commerce Incoming Telegram, April 1994, *supra* note 23; U.S. Department of Commerce Incoming Telegram, March 1994, *supra* note 23;

[28] U.S. Department of Commerce Incoming Telegram, Sept. 1994, Subject: Egyptian Cellular Project: Update.

[29] U.S. Department of Commerce, Incoming Telegram, June 1994, Subject: Colombian Economic Highlights 6/18-24/94; "Colombia Grants Licenses to Northern Telecom, Millicom," *Global Telecom Rep't,* June 13, 1994; "New Cellular Licensees Compete for Customers in Lucrative Market," *Lagniappe Letter,* May 27, 1994.

[30] U.S. Department of Commerce Incoming Telegram, March 1994, Subject: Colombia: Last Three Winners for Cellular Service; U.S. Department of Commerce Incoming Telegram, Jan. 25, 1994, Subject: Colombia: Winners of the Bid for Cellular Services; U.S. Department of Commerce Incoming Telegram, Jan. 24, 1994, Subject: Cellular Telephone Bid Public Meeting Schedules.

[31] The original bill applied to all telecommunications, whereas the second applied only to "value added services," a category into which cellular is considered to fall. Telephone conversation with Ed Zarnicke, *Latin Am. Telecom Reporter,* Feb. 3, 1995.

[32] In December 1992, Bill 230, which became Law 37, was passed by the Colombian Senate and House of Representatives. "Colombian Cellular Law Approved," *Latin Am. Telecom Reporter,* Jan 15, 1993, p. 4. It was subsequently signed by the President and so enacted into law. *See* Lisa Sedelnik, "Dialing for dollars; privatization of Latin America's telecommunications sector," *Latin-Finance,* March 1994.

[33] U.S. Department of Commerce, "Summary of Recent International Cellular Developments," 11/18/93, p. 4; "Cellular Concession," *Latin Am. Telecom Rep't*, April 1, 1993; "Colombia Cellular Law Approved," *Latin Am. Telecom Rep't*, Jan. 15, 1993, p. 4.

[34] David Scanlan, "Colombia goes loco for cell phones," *The Gazette* (Montreal), June 17, 1995, p. C2.

[35] *E.g.*, Chris Torchia, "Laundering Money in a System of Silence," *L.A. Times*, Nov. 20, 1994, § A, p. 13.

[36] Vicki Sinden and Zbynek Loebl, "Telecommunications Law in the Czech and Slovak Federal Republic," in *Telecommunications Law in Europe, supra* note 15, at 37; "Cellular telephony comes at a premium in East European Markets," *FinTech Mobile Communications*, Sept. 24, 1992. The Czech and Slovak licenses and joint venture agreements are identical. Conversation with Robert VanBrunt, *supra* Ch. 2, note 24.

[37] Joint Venture Agreement Between Post and Telecommunications Administration of the Slovak Republic and Atlantic West B.V., § 5.1. Action on most matters is to be taken by majority vote, although some issues must be unanimously decided. *Id.*, §§ 5.1, 5.4. In case of deadlock, the chair may cast a special vote to finally resolve the matter. *Id.*, § 5.6.

[38] *Id.*, §§ 1.1(z), 6.1, and 6.2. Section 6.2 further provides that this appointment is subject to approval of the PT, which shall not be unreasonably withheld, and that " [t]he parties anticipate that the Managing Director during this time period will be a person who is an employee of AWBV or its Affiliates."

[39] *Id.*, § 6.3.

[40] Conversation with Robert VanBrunt, *supra* Ch. 2, note 24.

[41] "World Update, Czech Republic/Slovakia," *Boston Globe*, Oct. 23, 1994. The article observes that the Czech Republic "has flourished" while Slovakia "has not." It cites statistics on unemployment, foreign trade, tourism, and inflation and notes that economic reforms such as privatization have proceeded more quickly in the Czech Republic.

[42] Prospectus, Grupo Iusacell, S.A. de C.V., May 20, 1994, p. 16.

[43] "An Interview with Raul Zorilla Cosio," *Eastern Europe and Former Soviet Telecom Reporter*, Oct. 15, 1993. See Telecommunications Regulations of the Ministry of Communications and Transportation of the United Mexican States, Ch. 2, art. 13, Oct. 25, 1990.

[44] Article 12 of the Federal Law of Telecommunications, published in the Official Journal of the Federation, June 7, 1995.

[45] Iusacell Prospectus, *supra* note 42, at v.

[46] "Bell South Sells Interest in Mexican Cellular Company to Grupo Iusacell," *Advanced Wireless Communications*, Feb. 16, 1994.

[47] Iusacell Prospectus, *supra* note 42, at 2, 67; "CTC Official Sums Up Effects of Privatization," *Eastern Europe and Former Soviet Telecom Reporter*, Oct. 15, 1993.

[48] "Market Trends," *Latin Am. Telecom Rep't*, July 1, 1994. It is believed that Motorola has received waivers allowing it to exceed the 49% limit on foreign investment.

[49] Iusacell Prospectus, *supra* note 42, at 16, 20; "Company News; Bell Atlantic Completes Purchase of Iusacell Stake," *N.Y. Times*, Aug. 12, 1994, § D, p.3, col. 1.

[50] Pursuant to a shareholders' agreement, the following actions (with some exceptions) are among those that are subject to supermajority vote: issuance of capital stock or recapitalization or change in capital or capital structure or determination that the company requires additional financial support from the shareholders; acquisition of the company's securities by the company; declaration or payment of certain extraordinary dividends or distributions; change in the number of directors; termination or disposition of a line of business above a specified dollar amount; incurrence of material indebtedness; material sale or material acquisition of assets, business, or earning power; commencement of a new business or line of business, or expansion of an existing business outside of Mexico; sales or acquisitions of assets in excess of a stated amount; entering into or amending contracts or transactions for the benefit of the majority shareholders; entering into or approval of

any transaction or plan of merger, consolidation, dissolution, or liquidation; subjecting the license to certain liens; making a request or application to obtain, modify, or terminate any license; amendment of the company's organizational structure or governing documents; changes in the company's officers or their remuneration; appointment or removal of the company's independent accountants or approval of financial statements; entering into material contracts; making tax elections or tax accounting methods or agreements with the government regarding taxes; approval or modification of any annual or three year plan; approval of terms of a private or public offering of the company's securities; granting of general or limited powers of attorney; sharing of proprietary technology; making of gifts or donations in excess of a specified amount; institution, settlement, or default with respect to material litigation or litigation in which the majority shareholders have an interest adverse to the company; and binding the company to an agreement not to compete.

[51] Iusacell Prospectus, *supra* note 42, at 51.
[52] *Id.* at 21.
[53] "Grupo Iusacell Announces First Half Results," *PR Newswire Investorfax*, June 26, 1995. This is about one-half the United States growth rate for the same period. At the time, Mexico was suffering from economic problems due to the devaluation of the peso.
[54] 47 U.S.C. § 310(b). Section 310(b) applies to radio licenses; foreign ownership of or interests in other common carrier authorizations is regulated by FCC rules promulgated pursuant to Section 214 of the Act.
[55] Joe Flint, "The last word from Commerce: relax; National Telecommunications and Information Administration on broadcast, cable regulation," *Broadcasting*, Jan. 25, 1993.
[56] 47 C.F.R. § 22.5.
[57] Decision, In re Applications of Algreg Cellular Engineering, CC Docket No. 91-142, ¶¶ 79–82 (Rev. Bd. 1994). The *Algreg* situation was considered particularly deplorable by the FCC because the applicants purposely concealed the alien ownership interest and because violations of other commission cellular rules were involved. Ch. 12, *infra*. It is possible that in less egregious circumstances a lesser sanction might have been imposed.
[58] Report and Order, In the Matter of Market Entry and Regulation of Foreign-affiliated Entities, IB Docket No. 95-22, FCC 95-475, at ¶¶ 1, 197–219 (1995).
[59] Declaratory Ruling and Order, In the Matter of Sprint Corporation Petition for Declaratory Ruling Concerning Section 310(b)(4) and (d) and the Public Interest Requirements of the Communications Act of 1934, as amended, FCC 95-498, Docket No. I-S-P-95-002 (released Jan. 11, 1996).
[60] John Van, "High-Tech Bet; Cellular's Success Makes New Technology Seem a Sure Thing," *Chicago Tribune*, Dec. 5, 1994, Business section, p. 1; First Report, In the Matter of Implementation of Section 6002(B) of the Omnibus Budget Reconciliation Act of 1993, Annual Report and Analysis of Competitive Market Conditions with Respect to Commercial Mobile Services, ¶ 3 (1995).
[61] Kenneth Kiplagat, "Fortress Europe and Africa Under the Lomé Convention," *N.C.J. Int'l & Comm. Reg. 589*, nn. 34 and accompanying text, 1993.
[62] Neocolonialism is described as a process by which developed countries seek to ensure that the ties that bound their colonies to them continue when those countries become independent. *Id.*
[63] Comments of Jean-Pierre Chamoux, Droit et Informatique, Paris, at Columbia Institute for Tele-Information seminar on International Ventures, Dec. 2, 1994.
[64] *Id.*
[65] David E. Sanger, "How Washington Inc. Makes a Sale," *N.Y. Times*, Feb. 19, 1995, § 3, p. 1.
[66] The failure of some joint ventures in China to reinvest profits within China was proffered as a reason for a recent proposal requiring the Chinese side of any joint venture between state enterprises and foreign companies to hold a stake of at least 51%. Kathy Chen, "China Drafts Measures to Limit Foreign Investment in State Assets," *Wall St. J.*, Nov. 18, 1994, p. A14. That same article notes, however, that the proposal may also be an effort by some government officials to maintain their clout in the Chinese economy.

[67] David Wessel, "IMF Urges Developing Nations to Study Controls on Inflows of Foreign Capital," *Wall St. J.,* Aug. 22, 1995, p. A2, col. 2.

[68] Comments of Marie-Monique Steckel, President, France Telecom North, at Columbia Institute for Tele-Information seminar on International Ventures, Dec. 2, 1994.

[69] John Rossant, "Italy: Berlusconi May Have Ruined His Best Chance for Change," *Business Week,* Aug. 1, 1994, p. 43; "Infrastructure: More Than US $500 Billion to be Spent on Projects In Hong Kong and China in the Next Ten Years," *Institutional Investor,* July, 1994, p. S8.

[70] These requirements were used by the government of New Zealand when it privatized its telephone company by selling it to Bell Atlantic and Ameritech. Similarly, while the Israeli government allowed 80% foreign ownership of its second cellular licensee, it imposed the condition that a majority of the board of directors be Israelis permanently residing in Israel. "Israel Issues Tender for Second Cellular License," *Mobile Phone News,* Nov. 22, 1993.

[71] For example, Atlantic West B.V. contributed cash of $7.667 million for its 49% share of Eurotel Bratislava's licenses. The licenses were contributed by the PT, which was then "deemed" to have contributed $7.980 million for its 51% share of the licenses. Exhibits B1, B2, and C to the Joint Venture Agreement.

[72] For example, the government of Taiwan required domestic bidders on a GSM license to have a qualified foreign partner. The stated reason was to advance completion of the cellular system. United States Department of Commerce, Summary of Recent International Cellular Developments, Updated 2/26/94, p. 6.

CHAPTER 5
▼▼▼

NUMBER OF LICENSES

5.1 INTRODUCTION

A government that wishes to issue authorizations to provide cellular service must confront the question of how many licenses it should provide. If only one license is issued, a monopoly will exist—except to the extent that cellular service competes with other products such as landline telephone service and paging (or, over time, satellite systems). When a monopoly is created, more regulation may be needed than when competition exists. Regulation carries with it certain costs. If more than one license is issued, then the discipline of market competition can substitute in part for regulation. However, each government must ask whether its economy can support more than one cellular provider.

5.2 REGULATION OF MONOPOLIES

The basic premise of public enterprise tariff policy is that "in the absence of good reasons to the contrary, prices charged should reflect the costs of providing services" [1]. Thus, a regulatory body must determine the costs of the telecommunications provider and set a fair rate of return for the business. By setting prices that reflect cost plus a fair rate of return, regulators attempt to ensure that the monopoly provider does not abuse its market power by setting excessively high prices [2].

This approach, known as *rate of return* regulation, encounters several difficulties. First, it is not always easy to determine what the telecommunications provider's "cost" is. Historical cost, being backward looking, is deemed an inappropriate measure. Accordingly, regulators use forward-looking costs to maximize economic efficiency. This naturally involves predictive estimations that may be difficult to define and measure. The problem is especially pronounced in the area of wireless communications, due to the rapid rate of technological change. Other problems involved in monopoly regulation are the difficulty in forecasting demand and the need to assign prices to specific services, or telecommunications *outputs* [3].

One of the other problems of rate of return regulation is that it provides the telecommunications operator with no incentive to reduce costs [4]. This can be mitigated by the use of a methodology called *price cap* regulation. Price cap regulation allows the operator to raise prices by an amount that does not exceed expected inflation minus some estimate of the extent to which the expected gain in productivity in the telecommunications sector exceeds the expected gain in productivity in the economy as a whole. However, one must still calculate the expected gain in telecommunications productivity. Moreover, price cap regulation may still provide an incentive to reduce the quality of service [5]. When a nonmonopoly provider reduces service quality it risks losing its customers to the competition.

Another problem with relying on regulation rather than market competition is what has been referred to as *X-efficiency* [6].

> The concept incorporates the ideas of motivation, incentive, and skill or, on the other hand, inertia, apathy, and ineptitude into the economic analysis of efficiency...The basic X-efficiency hypothesis is that neither firms nor individuals are always as productive as they could be, and, as a result, costs are not always kept to their minimum...All other things being equal, X-efficiency will be higher in a competitive than in a monopoly environment because each environment has different incentive characteristics.

Finally, any regulatory scheme has its administrative costs. Regulators must be hired and trained. Regulated entities must maintain organizations to keep cost records in the form and manner required by the government and to present rate cases. Appropriate changes in pricing may be delayed by the need to undergo administrative processes. The regulatory process may be subject to political pressures of the unhelpful sort and the interests of bureaucrats who desire to keep their jobs regardless of whether they are producing a benefit or detriment to society.

5.3 COMPETITION

Some of the problems associated with regulation of monopolies can be alleviated by the introduction of competition. Where a monopoly no longer exists, the regulatory structure required to avoid the ill effects of monopolization need no longer be put in place. Competition in telecommunications has been widely credited for growth in the sector, particularly in cellular [7]. For example, one study of 26 countries that had cellular systems in place as of 1987 found that for the period from 1987–1994, countries that allowed competition had much higher penetration rates than those with a single operator [8]. The Commission of the European Union has said that it will use its powers under Article 90 of the of the Treaty of Rome to force all EU nations to have awarded at least two mobile telephone service operating licenses by October 1996 [9]. (For a description of the EU's legal structure, see Section 7.2.)

Economists usually defend reliance on the market as a governance structure on the grounds that private sector resource allocation decisions will generally promote both allocative and technical efficiency [10]. "Allocative efficiency implies that real output cannot be increased by shifting resources from one activity to another. Technical efficiency assumes that in all activities being undertaken the value of the output is being maximized for any given expenditure on inputs" [11].

Empirical evidence on these theoretical propositions is mixed; however, governments have found qualitative evidence persuasive in some cases. They believe that competition promotes technological innovation and enhances responsiveness to customer needs (e.g., by creating new pricing and packaging of services). It has been argued that this is less important in developing countries, where demand is primarily for very basic services. However, even if advanced services are not in demand, technological innovation is important in developing countries because new technology can mean increased capacity [12].

Although one may start with the proposition that competition is generally to be preferred over regulated monopolization, one does not end there. Economic efficiency is a means to achieve social good, not an end in itself. Even in a competitive telecommunications market, regulators can play an important role: by ensuring universal access to telecommunications services and interconnection among networks, for example [13].

Moreover, adopting the competitive approach means that one forgoes the economies of scale that can be realized by a monopoly. Large economies of scale exist in industries in which efficient, low-cost production can be achieved only if producers are extremely large both absolutely and in relation to the market so that the firm's average cost schedule declines over a wide range of output. Historically, telecommunications has been perceived as having significant economies of scale [14]. However, there is some evidence that economies of scale no longer justify telecommunications monopolies. It is contended that the cost advantages of a monopoly are not as great as is sometimes presumed: due to technological change, telecommunications service providers can increasingly enter the industry profitably at a smaller scale than in the past. Moreover, the slower rate of technological change associated with monopolies may offset any advantages accruing from economies of scale [15].

Another potential advantage of issuing a nationwide monopoly license that may be lost when more than one operator is authorized is maximization of license fees paid to the government. The United States, which in the 1980s issued two cellular licenses for each of hundreds of metropolitan and rural service areas, subsequently decided to auction additional spectrum for PCS. The auctioned areas—called basic trading areas and major trading areas—are larger than the cellular territories for which licenses were granted in the 1980s [16]. For purposes of PCS, the United States is being divided into 51 major trading areas and 493 basic trading areas, with a total of 2,074 licenses (six in any one location) [17]. But as one FCC official noted, "Our primary goal is to create competition and economic growth, not to raise revenue...If revenue had been our goal, we could have gotten a lot more money by auctioning a nationwide monopoly" [18].

Where the political and economic situation of a country permits it, competition seems more likely than monopoly regulation to produce a better telecommunications system [19]. This conclusion raises the questions of whether more than one license can be supported by a country's economy and, if so, how many more than one are optimal.

5.4 OPTIMAL NUMBER OF CELLULAR LICENSES

How many cellular licenses should a particular government grant? The decision is complicated, because more does not necessarily mean better. Two licensees will provide for the competition (and attendant social benefit) that is absent when a single licensee has a monopoly. Three providers may provide for greater competition and benefits. But the presence of ten licensees will not necessarily provide for greater social benefit than three (or some number greater than three but less than ten).

As one analyst who has written on the subject has noted, one of the primary objectives of the government in granting cellular licenses is to maximize operators' incentive to invest [20]. As that author frames the question [21]:

> In any market there is in principle an optimum number of licenses that will achieve this objective: too few, and operators will prefer to cream-skim; too many, and the level and certainty of their projected financial returns will be too low to justify as much investment.

Thus, the operator's decision to deploy investment will depend on whether it can recover that investment through payments by its customers, or whether it will not be able to obtain enough subscribers because of the number of competitors in the market. Consequently, the appropriate number of licenses for a market depends in large part on how many potential cellular customers the population includes.

One would normally expect that the higher the per capita income of a country's citizens, the greater the number of potential cellular subscribers there are, and thus the more cellular licenses that can be supported. Other factors contributing to the success of cellular include high population density, small geographic area, and the presence of industries—such as construction and transportation—with high demand for cellular service [22]. The existence of highways is also a positive sign, presaging a potential for mobile use.

However, there are some surprising exceptions to the use of income as a guide to the success of cellular. For example, in 1994, Sri Lanka authorized its fourth cellular carrier [23]. In 1989, Celltel, a Swedish, U.S., and Sri Lankan joint venture, began the first cellular operations in the country with a TACS 900 system [24]. A second operator, Lanka Cellular, jointly owned by Singapore Telecom International and Capital Development and Investment Company of Sri Lanka, turned up another TACS system in January 1993 [25]. AMPS cellular service is now provided by a third operator, Mobitel, a joint venture between Australia's state-owned overseas telecommunications carrier and Sri Lanka Telecom, which also entered the market in 1993 [26]. The fourth cellular operator in Sri Lanka is Malaysian Telecom [27].

It is not clear that Sri Lanka will be able to support four cellular carriers. The ITU classifies Sri Lanka as a "low income" country [28]. It has a population of more than 17 million people [29]. Nevertheless, several factors seem to have facilitated cellular's popularity in Sri Lanka. First, the country's landline system has a low teledensity rate: less than one line per hundred people [30]. In some areas of the country, waiting lists for landline service run into the years [31]. Thus, there is a clear need for cellular as a bypass technology. Second, cellular phones have become more affordable, at about $250–$400 per set (depending on the source of the estimate), since the government abolished its import duty of 40% on cellular phones [32]. Third, the number of cellular providers has led to competition on pricing [33]. As of late 1994, there were approximately 23,000 cellular subscribers in Sri Lanka [34], for a penetration rate of about .13%, or 13 subscribers for every 10,000 people [35].

It is possible for the government to avoid the question of how many cellular operators to license by permitting open entry. The theory is that the financial community is better qualified than the government to determine the capacity of the market for operators. Therefore, the argument goes, the government should simply issue a license to each potential operator that seeks one, and allow them to compete in the capital markets for funds to build and operate their systems [36]. Of course, if the market makes a mistake and provides funds to too many operators, important benefits have been lost in the meantime. The government will have lost the money it could have earned from selling the optimal number of licenses to begin with. It will have foregone the opportunity to award licenses only to the applicants that promise the most favorable schedule of rates. Moreover, some qualified operators may choose not to invest given the uncertainty of recovering their investment. The potential benefit of their expertise will thus be lost. And spectrum is not an unlimited resource; if rights to utilize it are granted to a cellular operator who then fails to obtain financing, its use in the interim is sacrificed [37].

5.5 PHASED LICENSING AND TEMPORARY MONOPOLIES

A compromise solution to the question of how many licenses are appropriate is to grant cellular licenses in phases and allow the first operator a temporary monopoly for a specified number of years [38]. Such an approach allows the initial provider to be assured of an opportunity to recoup its investment [39] while still maintaining the threat of eventual competition. This specter of competition should induce the first licensee to be concerned with providing a high quality of service in order to build a loyal customer base. Phased licensing may also force subsequent licensees to focus on developing innovative new services to differentiate themselves from the first provider. Finally, it gives the regulator an opportunity to judge whether the market can support more than one provider [40].

This was the approach taken by the Czech and Slovak Ministry in 1990. The joint ventures between the local PTs and Atlantic West B.V. were given an exclusivity period of five years to operate all wireless telephony networks and provide wireless telephony service in the Republics, subject to revocation of their licenses for failure to perform in accordance with the license terms. The right to use the 450-MHz band for cellular was made exclusive for 20 years. The Czech and Slovak PTs and AWBV also agreed that for a period of 20 years they would not, without the consent of the other party, participate in any wireless telephony networks in the licensed areas, except through the joint ventures [41].

Notes

[1] Saunders, *supra* Ch. 2, note 1, at 255.

[2] *Id.* at 259.

[3] *Id.* at 257–81, 286–96. Pricing policy must encourage efficient allocation of resources and should, in principle, also include costs or benefits that are external to the market. Ch. 4, note 4, *supra*.

[4] One indication of this is the number of employees that have been laid off from the regional Bell operating companies in the United States since those entities have become subject to competition. Leslie Cauley, "BellSouth Will Trim 9,000–11,000 Jobs In Addition to the 10,200 Already Set," *Wall St. J.*, May 19, 1995, p. A3, col. 1.

[5] Saunders, *supra* Ch. 2, note 1, at 296–98.

[6] *Id.* at 281.

[7] *E.g.,* "Latin American Cellular Market Ripe for Growth, Report Says," *Global Telecom Rep't,* March 7, 1994.

[8] International Telecommunication Union, World Telecommunication Development Report, 1995, at 111.

[9] "EU To Pave Way For Mobile Phone/Data Services," *Dow Jones News*, Nov. 27, 1995.

[10] Telecommunications Sector Reform in Asia: Working Papers 103, Peter Smith and Gregory Staple (eds.), 1994 [hereinafter "Working Papers"].

[11] *Id.* at 103–05.

[12] *Id.* at 121–25. As discussed in Ch. 3 *supra,* a developing country may wish to provide advanced services.

[13] *Id.* at 106, 108, 118–19, 125–26; Ch. 10, 13, *infra.*

[14] McConnell, *supra* Ch. 4, note 4, at 560–61.

[15] Working Papers, *supra* note 10, at 113.

[16] Many telephone companies that received original cellular grants received them for several contiguous areas. And some cellular operators have bought properties from the initial licensees in order to aggregate them into large territories.

[17] Bulletin, Bell Atlantic Mobile, Nov. 2, 1994.

[18] Mary Lu Carnevale, "Bidding for Cellular Licenses Is Expected to Be Uneven Across U.S.," *Wall St. J.*, Nov. 2, 1994, p. B13.

[19] Arguably, even if only one cellular license is issued, that operator does not have a monopoly to the extent that its service is a substitute for landline service. To that degree, the cellular operator competes with the landline operator.

[20] Simon Glynn, "How many cellular licenses should there be? The economic feasibility," *Telecommunications Policy*, March 1994, pp. 91–92.

[21] *Id.* at 92.

[22] Asian Cellular Markets, Vol. I, *supra* Ch. 3, note 42, at 22.

[23] Rahul Sharma, "Sri Lanka a new boom market for cellular phones," *Reuter Business Rep't*, Sept. 21, 1994; James Riley, "Asia-Pacific blazes a new trail in crucial industry dominated by the state-owned monopolies; Telecoms opening to private sector," *S. China Morning Post*, June 30, 1994, p. 14.

[24] Asian Cellular Markets, Vol. I, *supra* Ch. 3, note 42, at 212; The Hooi Ling, "Singapore Telecom Unit aims for bigger slice of market," *Business Times*, Aug. 5, 1993, p. 11. The U.S. and Swedish ownership is in the hands of Millicom International SA, a multinational cellular company. Millicom International Cellular SA Annual Report and Accounts 1993 at pp. 1, 9.

[25] Asian Cellular Markets, Vol. I, *supra* Ch. 3, note 42, at 212; "Singapore Telecom unit aims for bigger slice of market," *supra* note 24. The IFC has made an investment in this system. "IFC Invests In Sri Lanka Cellular System," *Mobile Phone News*, March 1, 1993.

[26] Minoli de Soysa, "Sri Lanka's Growth Sparks Cellular Phone Boom," *Reuter Library Rep't*, July 23, 1993; "OTC wins in Sri Lanka," *FinTech Mobile Communications*, Dec. 17, 1992; "OTC to Provide Cellular System in Sri Lanka," *Exchange*, Dec. 11, 1992.

[27] "Sri Lanka a new boom market for cellular phones," *supra* note 23.

[28] This is the lowest category on the ITU's scale. In its annual report, the majority owner of Celltel listed gross domestic product per capita in Sri Lanka as $458. Millicom International SA Annual Report and Accounts, 1993, at p. 9.

[29] "OTC to Provide Cellular System in Sri Lanka," *supra* note 26.

[30] "Sri Lanka a new boom market for cellular phones," *supra* note 23.

[31] *Id.*

[32] *Id;* "Sri Lanka's Growth Sparks Cellular Phone Boom," *supra* note 26, However, the Sri Lankan government subsequently reimposed the duty on cellular phones, which is expected to slow the growth rate of cellular drastically. Gaston de Rosayro, *S. China Morning Post*, Jan. 28, 1995, Business Section, p. 4.

[33] "Sri Lanka a new boom market for cellular phones," *supra* note 23; "Sri Lanka's Growth Sparks Cellular Phone Boom," *supra* note 26. In its annual report, the majority owner of Celltel described two special price plans that it offered in response to "aggressive pricing adopted by one of its competitors..." Millicom International SA Annual Report and Accounts 1993, at p. 14. However, not everyone agrees with the argument that competition in cellular has brought affordable phone service to Sri Lanka. Pratap Chatterjee, "Development: World Bank Proposals No Good for Poor, Critics Say," *Inter Press Service*, June 19, 1994.

[34] "Sri Lanka a new boom market for cellular phones," *supra* note 23. Another source reported about 25,000 cellular subscribers. "ITU report commends Sri Lanka's telecom progress," *Xinhua News Agency*, Aug. 9, 1994.

[35] The majority owner of Celltel, reporting global operating profit of $7.3 million from cellular operations for the first half of 1994, stated that Sri Lanka was one of the four largest contributors to operating profits. "Millicom International Cellular S.A. Interim Report January–June 1994," *PR Newswire*, Aug. 8, 1994. Financial results were not reported for each country's operations.

[36] Glynn, *supra* note 20, at 93.

[37] See generally, Stephen Duthie, "Malaysian Firm Enters Crowded Phone Market," *Wall St. J.*, Dec. 23, 1994, p. B5A, col. 3. This article reports that a seventh mobile phone operator will be entering the Malaysian market in 1995 and questions whether that number of operators, using three mobile technologies, can survive. It notes the potential problems of overcapacity, redundancy, and poor commercial returns. It further points out the inability of the operators to interact due to differing technologies.

[38] The monopoly might expire earlier if the operator's subscriber base reaches a minimum number of customers. United States Department of Commerce, Summary of Recent Cellular Developments, Updated 11/18/93, at pp. 11, 19 (discussing provisions in Israeli and Turkish licenses).

[39] See "EU to Free Telecom Networks," *Wall St. J.*, Nov. 18, 1994, p. A14.

[40] Glynn, *supra* note 20, at 94.

[41] Joint Venture Agreement, *supra* Ch. 4, note 37 at §§ 17.1, 17.2; Exhibit I1 to Joint Venture Agreement, at §§ 3.1, 3.4.

CHAPTER 6
▼▼▼

LENGTH AND GEOGRAPHIC SCOPE OF LICENSES

6.1 TERM

The term of a license is important to a potential operator for obvious reasons. However, it can be deceptive to look only at the number of years for which a license is granted. Of equal significance is the question of whether the license is renewable and, if it is, what standards the operator must meet to be awarded renewal.

The initial term of years for which cellular licenses have been granted varies widely. However, most run from 10 to 20 years. Renewal periods may be for the same number of years as the initial term, or for a different number of years.

The license granted by Mexico to the predecessor of Iusacell is unusual in that the initial term was 50 years. The 1957 concession provides that at the end of the 50-year period the term can be extended "if the concessionaire has complied with the conditions of his concession title and accepts the new conditions established by the Federal Government" [1]. Without knowing what new conditions might be imposed, it is impossible to determine whether the renewal provision has any value. Cellular licenses granted to other cellular companies that were later acquired by Iusacell were assigned terms of 20 years, commencing in 1990 [2]. Article 19 of the 1995 Mexican Federal Law of Telecommunications sets an original term of up to 20 years and allows renewals for terms equal to the original.

In the United States, cellular licenses are granted for a term of 10 years, as are licenses for PCS. The renewal expectancy for each is characterized by the FCC as "high," and the FCC has adopted parallel renewal standards for PCS and cellular [3]. The FCC has established a detailed two-stage process for cellular license renewals [4]. If a licensee meets certain standards, it will be awarded a "renewal expectancy." If the expectancy is denied, a comparative hearing will be held to determine whether the incumbent or another applicant should be awarded the license.

Specifically, the process works as follows. Before the license expires, the incumbent and anyone who wishes to compete with the incumbent for the license will file their applications. If competing applications are filed, the FCC will issue a public notice, and the incumbent will then file its case for a renewal expectancy. Even if no competing applications are filed, interested parties will be allowed to file petitions against the incumbent's renewal expectancy [5]. To be entitled to a renewal expectancy, the incumbent must show two things. The first is that it has substantially used its spectrum for its intended purpose. For its performance to be "substantial," it must be "sound, favorable and substantially above a level of mediocre service, which might minimally justify renewal" [6]. The second is that it has substantially complied with applicable FCC rules and policies and the Communications Act [7]. "At a minimum" the incumbent's showing of why it believes it is entitled to a renewal expectancy must include the following [8]:

1. A description of its current service area in terms of geographic coverage and population served, as well as the system's ability to accommodate the needs of roamers;
2. An explanation of its record of expansion, including a timetable for the construction of new cell sites to meet changes in demand for cellular radio service;
3. A description of its investments in its cellular system.

If the renewal expectancy is granted and the incumbent is found to meet the "basic qualifications" for a cellular license, the proceeding will terminate and the license will be renewed. If the expectancy is denied but the incumbent meets the basic qualifications, the proceeding will move to step two. At that point, the FCC will decide whether the competing applicants possess the requisite basic qualifications. If they do, the FCC will then conduct an expedited proceeding to compare qualifications and determine which candidate should be awarded the license [9].

The FCC articulated three reasons why a high renewal expectancy is in the public interest. First, it will encourage investment in cellular facilities. Second, it will avoid the risk of replacing an acceptable service provider with an inferior one, based on unproven promises. Third, it will avoid disruption of cellular radio service [10]. Although the FCC has sought to avoid saying that as long as the cellular provider substantially does its job, the term of its license is effectively perpetual, the iference can be drawn [11].

The first filings for renewal of cellular licenses were made in 1993 and subsequent filings are being made on a rolling basis. No competing applications had been filed as of March 1, 1995. On January 23, 1995, the FCC granted the renewal applications of all those carriers filing in the first round (those in the top 30 markets) except for those against which a petition to deny was filed.

Within the European Union (EU) there has been a wide range of license terms. The practices of the various countries on this subject were summarized in a report commissioned by the EU as part of its effort to harmonize the laws of its member states regarding mobile communications [12]. The shortest term was 5 years, in Denmark. The longest was 25, in the United Kingdom. Several countries granted 15-year licenses. Ten and 20-year terms were also used by some member states. In some countries the license term was indefinite, although this was the case only for public licenses—that is, where the local telephone company was given the right to provide cellular as well as landline service [13]. In its subsequent Mobile Green Paper, the EU did not mandate a specific term or renewal standard to be adopted by its members. Rather, it stated that "[i]n order to foster innovation of systems and services, and to ensure in particular efficient use of frequencies, the duration of licences should be based on the period required to pay back investment on reasonable terms" [14].

The licenses issued to Eurotel Prague and Eurotel Bratislava by the Ministry of the Czech and Slovak Republics were for initial terms of 20 years [15]. The licensees were also given the right to apply for an extension term of 5 years. The extension [16]:

> shall be granted within thirty days of the application, unless it is demonstrated by appropriate evidence that Licensee, during the term of the License taken as a whole:
> (a) has not substantially used the License for the purpose of providing cellular telecommunications services to customers; or
> (b) has not substantially complied with the reasonable applicable rules and policies of the Ministry relating to cellular telecommunications as stated in this License.

When the International Bank for Reconstruction and Development (World Bank) undertook to assist the Kingdom of Jordan in preparing an invitation for tenders for a GSM cellular system, the draft that resulted stipulated a license term of 15 years, to be extended for 5 years at a time, subject to satisfactory performance [17]. The License "shall always be renewed if the Licensee has operated successfully and in accordance with the laws and the License and if there are no other reasons to refuse the renewal after successful negotiations" [18]. The last part of that provision regarding renewal creates a substantial amount of uncertainty for the investor.

6.2 GEOGRAPHIC SCOPE

A country may decide to grant cellular licenses that are nationwide in scope, or that cover less than the entire country [19]. As might be expected, national service areas tend to be more prevalent in geographically smaller countries. However, some countries have issued both nationwide and regional licenses. Moreover, large countries that have issued licenses for less than national areas have often seen cellular operators aggregate licenses by purchasing them from the initial licensees in order to provide their customers with larger local service areas and to take advantage of economies of scale.

In issuing both its cellular and PCS licenses, the U.S. FCC decided to divide the country into a large number of markets. However, the market structure for PCS, for which auctions started in 1994, is different from the market structure for cellular, for which licenses were issued earlier. PCS licenses will cover larger service areas than cellular licenses. Moreover, PCS bidding to date has demonstrated a pattern of "clustering." That is, bidders are often attempting to buy licenses for contiguous areas so as to aggregate them into a single large service territory.

Cellular markets 1 through 305 are metropolitan statistical areas, market 306 is the Gulf of Mexico, and markets 307 through 734 are rural service areas [20]. The A side license was awarded to a "nonwireline" applicant and the B side license was given to an affiliate of the local landline telephone company or companies [21]. At the time the scheme for cellular licenses was first being developed, the local landline company in most areas (except those parts of the country served by independent telephone companies) was AT&T. The FCC was attempting to expedite provision of service to the largest urban areas, where there was a shortage of traditional mobile frequencies. The Department of Commerce had already designated standard metropolitan statistical areas based on census numbers and the FCC used these designations, with some adjustments. The RSA division occurred later, in 1986. The FCC's goals were to have clearly defined areas, big enough to be economically viable for multiple licensees and to avoid "cream skimming," whereby applicants would seek to serve only small, more lucrative areas [22].

This division of the country into such a large number of different cellular service areas did not survive, however. Large telephone companies, such as the regional Bell holding companies that succeeded to the cellular licenses of AT&T [23], effectively received large service territories because of their landline presence in many contiguous markets. Nonwireline operators, such as McCaw Cellular, began to aggregate A side properties to compete with the Bell companies, and the Bell companies in turn began to purchase A side properties to expand their service territories [24]. It appears that, like virtually everyone else, the FCC of the early 1980s underestimated the demand for cellular service. A government making a decision on geographic scope today has the benefit of knowing what the FCC did not at that time: due to demand for wireless services, large companies are willing to invest the capital needed to take advantage of the economies of scale that can be gained from using one organization to serve a large region, or, in many cases, an entire country.

When it issued its Notice of Proposed Rulemaking on PCS, the FCC noted this consolidation of cellular markets and the amount of money that had been spent in transaction costs—such as brokers' fees—incurred in connection with the aggregation process. Accordingly, it proposed a range of service area options for the new personal communications services, all of them involving fewer areas than there were for cellular: (1) the 487 "Basic Trading Areas" defined in the Rand McNally *Commercial Atlas and Marketing Guide*, plus Puerto Rico; (2) the 47 Rand McNally "Major Trading Areas" (composed of multiple BTAs); (3) the 194 telephone local access and transport areas (LATAs), or areas in which a local Bell telephone company were permitted to carry a call without having to transmit the call by way of an interexchange carrier such as AT&T; and (4) the nation. It also suggested that if given authority by Congress, it might allow the size of the service area to be determined by competitive bidding [25]. After reviewing public comment, the FCC concluded that a combination of BTAs and MTAs would be appropriate, given that they are "based on the natural flow of commerce," and "have been determined after an intensive study of such factors as physiography, population distribution, newspaper circulation, economic activities, highway facilities, railroad service, suburban transportation, and field reports of experienced analysts" [26]. Another factor supporting the use of BTAs is that the FCC wished to encourage ownership by women and minority-owned and small businesses, which might not have access to the amounts of capital needed to acquire licenses for larger territories.

The Mexican government divided the country into nine cellular service regions. However, Telcel, the affiliate of the national telephone company, Telmex, effectively received a national license because it was given a license in each region. As in the United States, aggregation of licenses in the aftermarket has occurred [27].

The Colombian licenses discussed above are also regional licenses. The country was divided into three regions, for both the private and mixed licenses. However, each of the seven prequalified bidders on the B (private) side licenses actually bid on all three regions. One bidder received the licenses for two regions [28].

National licenses are often granted from the start in smaller countries. Each of the EU countries has issued national licenses [29]. Although the Czechoslovak licenses divided what was at the time one country into two service areas, the analog licenses were granted to the same principals; they have since become national licenses by virtue of the decision to become two countries. Both the Ugandan and Jordanian GSM licenses discussed in prior chapters are national in scope. In light of the experience of larger countries that have seen millions of dollars expended in transaction costs in aggregation of cellular licenses in order to take advantage of economies of scale, the decisions by smaller countries to grant national licenses are easy to understand.

Notes

[1] Article Fourteenth of the 1957 Concession granted April 1, 1957 by Ministry of Communications and Public Works to Servicio Organizado Secretarial, S.A.

[2] Iusacell Prospectus, *supra* Ch. 4, note 42, at p. 67.

[3] Second Report and Order, In the Matter of Amendment of the Commission's Rules to Establish New Personal Communications Services, ¶¶ 130–31, GEN Docket No. 90-314 (1993).

[4] Memorandum Opinion and Order on Reconsideration, In the Matter of Amendment of Part 22 of the Commission's Rules Relating to License Renewals in the Domestic Public Cellular Radio Telecommunications Service, 8 F.C.C. Rcd 2834, 72 R.R.2d 553 (1993). The time period for filings under this decision were modified slightly in Memorandum Opinion and Order on Further Reconsideration, 9 F.C.C. Rcd 4487, 75 R.R.2d 776 (1994). The regulations promulgated under these decisions are now found at 47 C.F.R. §§22.935 and 22.940.

[5] Memorandum Opinion and Order on Reconsideration, In the Matter of Amendment of Part 22 of the Commission's Rules Relating to License Renewals in the Domestic Public Cellular Radio Telecommunications Service, *supra* note 4, at ¶ 24.

[6] *Id.* at ¶¶ 3, 8.

[7] *Id.* at ¶ 4. In determining whether an incumbent has complied with FCC rules and policies and the Communications Act, the FCC will also examine the misconduct of certain persons affiliated with the licensee. *Id.*, n. 6.

[8] *Id.* at ¶ 10 & n. 7.

[9] *Id.* at ¶ 25. Basic qualifications include such things as the applicant's financial qualifications to provide service, whether it meets the alien ownership rules, and whether it has engaged in non-FCC misconduct.

[10] *Id.* at ¶ 13.

[11] For an example of a country granting a perpetual license, see Peter Bate, "GTE, ATT Team Up to Sweep Argentine Cellular Tender," *Reuter Business Rep't*, Nov. 5, 1993.

[12] Licensing and Declaration Procedures for Mobile Communications in Member States of the European Community (Final Report, Aug. 1993)[hereinafter "Licensing and Declaration Procedures"]. The report was prepared by KPMG Peat Marwick of London and Stanbrook and Hooper of Brussels.

[13] *Id.* at pp. 24–25, 32. The Danish license appeared to be automatically renewable for another five-year term, although the additional term was not legally guaranteed. *Id.* at Annex One, p. 16. The countries studied were Belgium, Denmark, France, Germany, Greece, Ireland, Italy, Luxembourg, the Netherlands, Portugal, Spain, and the United Kingdom.

[14] *Overview, supra* Ch. 3, note 56, at p. 8, § I, ¶ 6.

[15] Joint Venture Agreement Between Post and Telecommunications Administration of the Slovak Republic and Atlantic West, B.V., Exhibit I2, ¶ D/1.

[16] *Id.*, ¶ D/2.

[17] The Hashemite Kingdom of Jordan, Ministry of Post and Telecommunications, The Telecommunications Corporation, Request for Tenders (TCC 24/93), Concession to Provide GSM Services, §§ 1.2.4, 3.3.

[18] *Id.* at § 3.19.

[19] In discussing the scope of licenses, this section refers to the area that the operator has the right to serve. An operator's obligation to serve the area, which is another matter, is discussed in Ch. 10, *infra*. Thus, for example, simply because a nationwide license has been authorized does not mean that the entire country necessarily has access to cellular service.

[20] Cellular Telephone Industry Association, "State of the Cellular Industry," 1992.

[21] The A side license operates on the 824- to 834- and 869- to 879-MHz bands and the B side license operates on the 835- to 845- and 880- to 890-MHz bands. Some additional spectrum is provided for control channels, which are used to convey information for call setup. Balston & Macario, *supra* Ch. 1, note 1, at 49–50.

[22] Report and Order, In the Matter of An Inquiry Into the Use of the Bands 825–845 MHz and 870–890 MHz for Cellular Communications Systems; and Amendment of Parts 2 and 22 of the Commission's Rules Relative to Cellular Communications Systems, 89 F.C.C.2d 58, ¶¶ 61–63, 50

R.R.2d 1673 (1982); First Report and Order, In the Matter of Amendment of the Commission's Rules for Rural Cellular Service, 60 R.R.2d 1029, 1033–34 (1986).

[23] AT&T was required, effective January 1, 1984, to divest its local telephone companies. This was the result of a consent decree referred to as the "Modification of Final Judgment" that was entered in an antitrust case brought by the government against AT&T. AT&T kept its long-distance business. However, AT&T transferred its cellular licenses to the seven regional Bell holding companies (the parent companies of the local telephone companies that were spun off). Ironically, ten years later AT&T acquired McCaw Cellular in an effort to reconstruct the virtually national cellular license that it had passed on to the Bell companies.

[24] *E.g.,* Louis Kehoe, "Survey of Mobile Communications," *Fin'l Times,* Sept. 5, 1994. Because the A and B side licenses use different parts of the radio spectrum, licenses for the two sides are not completely fungible given the current state of technology.

[25] Notice of Proposed Rulemaking and Tentative Decision, In the Matter of Amendment of the Commission's Rules to Establish New Personal Communications Services, at ¶¶ 56–61, GEN Docket No. 90-314, ET Docket No. 92-100 (1992).

[26] Second Report and Order, In the Matter of Amendment of the Commission's Rules to Establish New Personal Communications Services, ¶ 73, GEN Docket No. 90-314 (1993).

[27] For example, Iusacell's cellular operations originated in Region 9, which includes Mexico City. It was granted the nonwireline cellular concession there in 1989. It was awarded 35% ownership of Regions 6 and 7 in 1990. By the end of February, 1994, Iusacell had acquired ownership of 100% of the nonwireline cellular concessions in Regions 5, 6, and 7. Iusacell Prospectus, *supra* Ch. 4, note 42, at p. 22.

[28] U.S. Department of Commerce Incoming Telegram, Jan. 25, 1994, Subject: Colombia: Winners of the Bid for Cellular Services.

[29] Licensing and Declaration Procedures, *supra* note 12, Annex One.

CHAPTER 7

▼▼▼

THE GRANT PROCESS

7.1 INTRODUCTION

A number of processes can be employed to determine who should be awarded a cellular license. Possibilities used or considered to date have included the following:

- Automatically granting the right to provide cellular service to the existing landline telephone company (a "wireline set-aside");
- Lottery;
- Case-by-case ad hoc decisionmaking;
- Comparative qualitative evaluation of bids (a "beauty contest");
- Comparative hearings;
- Auction.

Some combination of the above (e.g., an auction among bidders who succeed in the initial beauty contest) may also be used [1].

The first three approaches—an automatic grant to the country's telephone company, ad hoc decisionmaking, and lotteries—have often been criticized. The criticism is that they are not well-designed to serve either the social aims that telecommunications systems are supposed to achieve or the goal of maximization of government revenue. Comparative methods and auctions have been easier to justify, although each has its strengths and flaws.

7.2 WIRELINE SET-ASIDE

The automatic grant of a cellular license to the local telephone company, also referred to as a "wireline set-aside," was used by the United States FCC to distribute cellular licenses in the 1980s. One of the two licenses in each service area was awarded to the local telephone company. Where there was more than one in the area, the license was awarded to a partnership of those companies or, if they could not reach full settlement, the one that won a lottery. As has been noted, that local telephone company, at the time of the FCC's decision, was in most cases AT&T. The FCC indicated in a 1981 decision that it felt that AT&T had demonstrated, through its activity in developing cellular technology, "that it possesses the resources and the expertise necessary to establish expeditiously cellular systems with nationwide capacity" [2]. It examined whether the role of the landline carriers should be limited on the grounds that if cellular could be considered a substitute for landline service, the telephone companies might have "a disincentive to fully develop cellular" [3]. It concluded that "[b]ecause of the likely unavailability in the near term of portable units that are truly substitutable for landline telephones and the inherent spectrum limitations on cellular systems [footnote omitted], we cannot conclude that wireline telephone service is within the relevant market [that is, the market for cellular service] in this proceeding" [4]. Technology today has made obsolete the idea that cellular is not substitutable for landline service in developing countries.

Most of the countries of the European Union—all except Greece—have also automatically granted cellular concessions to their landline telephone companies [5]. These companies frequently were government entities, although many have been or are being privatized. The report commissioned by the EU on mobile communications noted certain flaws with this set-aside policy. First, such vertical integration may result in unfair trade practices. These include cross-subsidization of competitive services by monopoly services and preferential terms for interconnection with the landline network being given to the telephone company's affiliate but not to its cellular competitor [6]. Second, a wireline set-aside does not serve important objectives of the regulator of cellular services: either to promote efficient use of spectrum and to realize a capital gain for the state (the arguments for auctions) or to encourage competition in mobile services and low end-user prices (the arguments for beauty contests or comparative hearings). Third, it may lead to perverse incentives when a competitor is being chosen for a second license: if the landline provider is a governmental entity or affiliate, the government might select a weak competitor so that the cellular provider affiliated with the government will succeed. Fourth, the telephone company may have too much of a head start on the competition if the grant process for the second license takes too long. The one advantage to a wireline set-aside that was noted by the EU Report was that it might entail certain operational and marketing economies of scope. That is, resources could be shared by the landline and cellular operations, and customers could be offered package deals that combine both types of service [7]. The question to be asked in any given situation is whether this gives the wireline carrier an unfair advantage over its competitors.

In the Green Paper on mobile communications for which the report referenced above was commissioned, the Commission of the European Communities (CEC or Commission) said that if a license is to be given to the landline telephone company, exclusive and special rights should nevertheless be abolished. That is, the phone company should not be given the only license and it should not be given a license or lasting particular advantage based on other than "objective, proportional, transparent, and non-discriminatory criteria" [8]. This finding was embodied in a subsequent communication to the European Parliament and Council of Ministers based on the public consultation that followed the Green Paper [9].

At this point, it may be useful to offer some background on the lawmaking process in the EU. The CEC consists of 17 commissioners, each responsible for one or more directorate-generals (DG). DG XIII is the principal DG responsible for telecommunications and formulates and monitors telecom policy. The CEC's legislative powers are generally circumscribed. One exception to this is its powers under Article 90 of the EC treaty, which the commission has a duty to apply. Article 90 provides that undertakings entrusted with the operation of "services of general economic interest or having the character of a revenue-producing monopoly" are subject to the treaty's rules regarding competition. The CEC takes the position that government-owned telecom operators fall within the ambit of Article 90. Other than with respect to its direct power under Article 90, the CEC submits proposals to the Council of Ministers, which is the prime law-making institution. The Council has the power to issue regulations, directives, and decisions (binding), as well as recommendations and opinions (nonbinding). Once proposals have passed the Council, they must be implemented by all member states if binding. The European Parliament has limited power compared to national governments, although the Council must consult with it [10].

7.3 LOTTERIES

Lotteries are often considered to be a quick, low-cost way of awarding licenses. When cellular licenses were awarded in the United States over a decade ago, nearly all of the nonwireline licenses (i.e., the A-side licenses, those not given to an affiliate of the local telephone company) for those areas other than the top 30 MSAs were generally awarded by lottery. Lotteries were also used for B-side licenses where there was more than one telephone company applicant. (On both the A and B sides, full settlements among competing applicants sometimes obviated the need for a lottery.) The FCC had planned to award the licenses by comparative hearing, but had a "significant backlog" of applications for a large number of markets. It hoped that the use of lotteries would speed the delivery of cellular service and save money [11].

The FCC considered using lotteries as a way to issue PCS licenses. It stated that they were superior to competitive hearings in terms of speed and expense [12]. Nevertheless, it did note the drawbacks of lotteries. As one study of the cellular lotteries that was cited by the FCC noted: "'Application mills' soon offered a standard completed application at modest cost. Because application costs were low and licenses valuable, the FCC received large numbers of speculative applications, imposing a major administrative cost" [13]. As discussed below, the auction method was eventually adopted for PCS.

Three primary arguments against lotteries have been cited by the International Telecommunications Union [14]:

- They transfer a "windfall" economic benefit to the lottery winner.
- The lottery winner's qualifications to operate a service may be questionable, even if some degree of prequalification has been required of participants in the lottery.
- They eliminate the regulator's ability to apply public-interest criteria to the choice of licensee.

The European Union's Mobile Green Paper found that "[l]otteries do not seem to guarantee the achievement of the criteria" that the EU believes should be used to evaluate a license award system. In particular, the Mobile Green Paper concluded that lotteries do not encourage "efficient use of frequencies, technical competence and financial resources" [15]. Public consultations following the Mobile Green Paper found "virtually no support" for lotteries [16].

7.4 AD HOC DECISIONMAKING

The ad hoc decisionmaking approach is often criticized because the result may be that the license is given to a licensee with the right "connections" to highly placed government officials [17] or to the one that influences the selection process by way of a "a 'hefty bribe' in the side pocket" [18]. Even if the process is not actually tainted, the absence of openness may lead to speculation that it is, which can be detrimental to the government's credibility and effectiveness.

Because of allegations of nepotism, the South Korean government was forced to withdraw the 1992 award of its second cellular license, thus delaying the introduction of competition by about two years. The first license had been given to Korean Mobile Telecommunications Corp. (KMT), a part of Korea Telecom, in 1984. A second license was issued in August, 1992, after bids for the license were evaluated [19]. However, the award immediately came under attack because the chairman of one of the companies in the winning consortium, Sunkyong, and Korea's then-President, Roh Tae Woo, were linked through the marriage of their children. Sunkyong declined the license—allegedly for fear it would be revoked anyway—and the decision to reaward it was postponed until after the presidential elections in 1993 [20]. The license was finally awarded in May 1994 to a consortium of U.S. telephone companies, a U.S. cellular equipment manufacturer, and Korean firms [21]. It is expected that KMT, the owner of the first cellular license, will continue to have a monopoly until as late as 1996, due to the delay in awarding the license as well as subsequent demands upon the consortium by the Korean government [22].

7.5 COMPARATIVE METHODS AND AUCTIONS

The remaining approaches—beauty contests, comparative hearings, and auctions—are all subject to legitimate debate on their merits. If comparative criteria are used, however, what seems to be most important is that the process be conducted according to rules that are known in advance to the prospective licensees and that are consistently and fairly applied.

The typical beauty contest comprises the following steps [23]:

- An initial public notice, issued by the regulator, defining the specific characteristics of the license(s) to be issued, the information required from bidders, and the evaluation process and criteria.
- Preparation and submission of bids by organizations seeking licenses.
- Evaluation by the regulator, which may or may not involve some form of structured "scoring" system.
- Selection of the winner(s) by the regulator.
- Issuing by the regulator of the decision, which may or may not be accompanied by a statement of the reasons for the choice with a greater or lesser degree of detail provided.

The ITU has stated that the advantages of the beauty contest are: it forces bidders to prepare detailed and informative bids, providing the regulator with a large amount of information on which to base a decision; and it may provide each bidder with an incentive to develop innovative services or pricing to differentiate its bid from others, if this is articulated by the government as a factor in the selection process. The disadvantage, according to the ITU, is that the process is costly, demanding significant resources from both the regulator and the bidders [24].

A straight beauty contest (that is, one that did not have an auction among prequalified bidders tacked on at the end of it) was used by Argentina and South Africa. It is instructive to examine the criteria that each used in evaluating bids. (Copies of the invitations to tender issued by those two governments appear in Appendices A and B to this book.)

Argentina awarded a nationwide (excluding Buenos Aires) license [25]. Argentina's primary concerns were coverage and competition [26]. To encourage this, the only license or revenue-based fees that were required to be paid to the Argentine government were one-half of one percent of revenues to reimburse the government for its costs of controlling, monitoring, and verifying cellular service providers [27] and a fee based on the number of subscribers in service [28]. The criteria articulated by the government as relevant to the judging were how quickly service could be provided, how broad coverage could be in terms of population and geography, interconnection and coordination with other telecommunications systems in the country, and quality of service [29]. The winning bidder promised to put a nationwide system of 700 cell sites in place within 30 days, significantly shorter than the normal time to build out a network of that size [30].

The standards set by the South African government for its national licenses were broader [31]. The first was proposed service: the qualities and facilities; the extent to which the choice of technology would lead to high volumes worldwide and low unit costs and competition; the extent to which the technology would support and foster trade and industry in South Africa; and the quality, facilities, reliability, and affordability of the service and appropriateness of the technology. The other criteria were financial resources, expertise, and experience of the operator and the extent to which know-how would be transferred to South Africa; the extent to which community service obligations (such as service to underdeveloped parts of the country) would be met [32]; ability to have the system operational quickly; a business plan and marketing strategy to secure high penetration and affordable rates; the cost of equipment and service; the local added value; job creation; extent of foreign equity; impact of foreign exchange; and the utilization of existing infrastructure.

In a comparative hearing, the process is similar to the beauty contest. In addition, evidence for and against the award of a license to a particular licensee is given orally by intervenors in hearings that are often public. However, regulatory officials who have had experience with comparative hearings argue that not only are they time consuming, they are not conducive to bringing useful information to light [33]. The FCC, which used comparative hearings to award licenses until 1984 [34], abandoned them because of their cost and delay. Years of litigation led to huge legal bills for the parties involved. The delay meant lost revenue for the licensee and years without service for the public. Existing service providers were charged with using the hearing process to retard their competitors' entry into the market. The FCC also found it difficult to formulate comparative criteria. Finally, it noted that if licenses are freely transferable, the licensee might not be the one that actually provides the service [35].

The general—but not unanimous—opinion seems to be that if a comparative procedure—hearings or a beauty contest—for awarding licenses is used, then openness, objectivity, and transparency lend an air of fairness that has a value in itself. According to the authors of the EU study on the issue, many in the industry felt that the procedure should exhibit the following characteristics [36]:

- All key criteria are published before bidding begins.

- All relevant factors, such as interconnection conditions and payments, frequency fees, and license payments are known before bidding begins.
- Criteria are based on maximizing customer benefit and choice.
- The promises made in the bidding process are incorporated into the license.
- An independent audit is available should the losing bidders request it.
- The selection process is open to judicial review.

The Directorate-General of Telecommunications of the European Union indicated its general agreement with the recommendation in favor of open, nondiscriminatory, and transparent procedures [37]. Public consultation on the issue found "overwhelming support from all sections of the industry for the use of competitive [nonauction] bidding" [38].

Nevertheless, as the EU study that preceded the Mobile Green Paper noted, it will not be easy to know all of the key criteria in advance, and following the published criteria exactly can lead to the wrong choice. Moreover, too much specificity in published criteria may encourage prospective operators to submit mechanical bids that slavishly follow the criteria and discourage innovation [39].

The merits and demerits of auctions have also been vigorously debated. Studies that have been used by the FCC and the EU provide good summaries of the positions.

The FCC sought comment on the use of auctions in its PCS rulemaking proceeding. A study cited by the FCC touted the following benefits of auctions: speed; low social cost (less engineering and legal time required to prepare applications than for comparative methods); low administrative cost; generation of information about value of spectrum that might be useful to the FCC in the future; and return to taxpayers for the value conferred in a license. One drawback noted was that it might be difficult for small bidders to obtain financing. The effect on ownership was found to be the same regardless of which method was used, assuming free transferability of licenses [40].

The FCC also examined issues relating to the design of auction procedures. It considered whether sealed or oral bidding should be adopted. It was suggested that sealed bidding is simpler to administer, less subject to collusion, and more likely to bring a higher price for the item being sold. Collusion is arguably more difficult under sealed bidding because the colluding parties run a greater risk of losing the bidding to a firm that reneges on its commitment to collude. Sealed bidding may bring a higher price because if there is only one bidder, the party will not be aware of that fact as it would in oral bidding. Therefore, a sealed bid process will compel the sole bidder to bid what it actually believes the spectrum is worth. However, if one can assume no collusion and multiple bidders, oral bidding is believed to have certain advantages. The concern is that in sealed bidding, the bidder who actually values the license the most may shade its bid too low because it will be trying to outbid the next highest bidder by as little as possible and may misjudge how much it needs to bid. In oral bidding one knows the amount of the competing bids [41].

Ultimately, the U.S. Congress authorized the FCC to use auctions to award licenses for new wireless services. Congress granted this authority in the 1993 Omnibus Budget Reconciliation Act. The Budget Act instructed the FCC that if it decided to use auctions, it was to devise a method to promote, among other things, the following objectives: development and rapid deployment of new technologies, products, and services for the benefit of the public, including those residing in rural areas, without administrative or judicial delay; economic opportunity and competition by avoiding excessive concentration of licenses and disseminating licenses among a wide variety of applicants, including small businesses, rural telephone companies, and businesses owned by members of minority groups and women; recovery for the public of a portion of the value of public spectrum resources and avoidance of unjust enrichment; and efficient and intensive use of spectrum [42]. The FCC, as a result, implemented a complicated system of multiphase, multiround electronic bidding that commenced on December 5, 1994 [43]. The first phase was completed on March 13, 1995, and winning bids totaled in excess of seven billion dollars [44].

The study of mobile licensing and declaration procedures commissioned by the EU to assist it in its formulation of the Mobile Green Paper expressed a preference for nonauction bidding. The study recognized the advantages of auctions as compared to lotteries. It also acknowledged a strong theoretical argument in favor of auctions: that in a perfect market the winner will be the one that can economically exploit the resource most effectively by achieving the greatest output for the least input [45]. But it questioned whether real-life markets are perfectly efficient. Moreover, it expressed the opinion that auctions are less socially desirable than a process that selects licensees via criteria that directly contribute to the maximization of consumer surplus, by the following [46]:

- Setting minimum conditions for social obligations such as quality of service and geographic coverage;
- Judging bids primarily on their proposed basket of tariffs together with a tariff control mechanism, (subject to satisfying the NRA [national regulatory agency] that the commercial plans of the lowest tariff bidder are feasible, and that bid is credible in terms of the organization's ability to provide the resources to support it);
- Making the tariff and other proposals, such as geographic coverage, conditions of the license and enforcing them through a firm regulatory regime, with graduated penalties for breaching license terms.

The primary argument against auctions identified by the study was the belief that if an operator is required to pay huge amounts for the right to use spectrum, it will be obliged to pass those costs on to end users in the form of higher prices [47]. However, the study did concede that under some circumstances the use of an auction is more appropriate. One that might be relevant for a developing country is where "a country has a very high need for short-term Treasury gain...[I]n that case it may be worth accepting the short-term gain despite reduced subscribers, less employment, and so less direct and indirect taxes" [48].

Although the consultants' study prepared for the EU discouraged the use of auctions, the Mobile Green Paper was less committal. It stated that whatever method is used, a priority should be "[maximizing] benefits for users (in particular, in terms of price and coverage)" [49]. It cautioned that "[r]eliance on auctions should not lead to an excessive transfer to the public budget or for other purposes to the detriment of low tariffs for the users" [50]. However, it stopped short of advising against the use of auctions.

The ITU has suggested a modified form of auction as a compromise. This approach would involve the selection of a "pre-qualified 'short-list' of the most qualified recipients of licenses through a simplified 'beauty contest,'" followed by an auction to select the actual licensee(s) [51].

The Jordanian request for tenders seems to have taken this sort of approach. Bidders must prequalify by meeting criteria for financial strength and cellular expertise and accepting the conditions imposed by the Jordanian government with respect to such items as coverage area and services to be offered [52]. Bids that satisfy these standards will then be evaluated for the financial benefits they promise the government. In the event that two or more bids offer about the same return to the government on a net present value basis, the bid with the lowest prices to users would be selected [53].

The 1995 Mexican Federal Law of Telecommunications also requires the Secretary of Communications and Transportation to establish a program to auction licenses to use spectrum for cellular, with prequalification of bidders. Bidders must fulfill certain requirements in order to participate in an auction: programs and commitments regarding the investment, coverage, and quality of services they propose to provide; a business plan; technical specifications for the project; and a favorable opinion from the Federal Commission on Competition. If the proposals submitted at the public auction do not guarantee the best conditions for providing service, or the Secretary does not consider the price obtained satisfactory, or the proposals do not comply with the requirements established for participation, the auction shall be declared null and a new auction conducted [54].

Either an auction or a comparative process—generally not involving hearings—seems to be the most common option selected by governments today. There is still a strong debate about whether an auction or a beauty contest is more effective. If government undertakes to select what it believes is the most qualified applicant (other than by strict economic efficiency qualifications), however, it would seem to be logical to prohibit the licensee from transferring its concession to another party that may be less qualified. If an auction is chosen, government may wish to set certain minimum qualifications for participation, possibly though a beauty contest.

Notes

[1] ITU Regulatory Colloquium No. 1: The Changing Role of Government in an Era of Deregulation xxi, 82–86, May 1993. Although the ITU does not seem to be using the term "beauty contest" in a pejorative sense, it has carried a negative connotation when used by others. The derogatory tone arises from the implication that the criteria in a beauty contest are always so subjective that they serve as a cover for an award of a license to the best lobbyist rather than to the best cellular operator. *See* Smith and Staple, *supra* Ch. 1, note 7, at 53. While this is sometimes the case (as is illustrated by the discussion of the South Korean cellular licensing process, *infra*), a process that involves qualitative criteria need not always be a farce. Accordingly, this paper uses the term "beauty contest" in the ITU (nonderisive) sense.

An additional option mentioned by the ITU report, grant of a franchise license by legislative action, seems unsuited to the selection of a specific cellular licensee. It might be appropriate when a collective license, covering an entire category of providers of service, is being granted. ITU Regulatory Colloquium No. 1: The Changing Role of Government in an Era of Deregulation at 82.

[2] Report and Order, *supra* Ch. 3, note 53, at ¶ 29 (1981). A number of parties challenged the wireline set-aside, but on reconsideration the FCC affirmed that this policy would best meet its goals of prompt initiation of service and development of a competitive market structure. Memorandum Opinion and Order on Reconsideration, 50 R.R. 2d 1073, 89 FCC 2d 58 (1982).

[3] *Id.* at ¶ 31.

[4] *Id.* at ¶ 32.

[5] Licensing and Declaration Procedures, *supra* Ch. 6, note 12, at pp. 24–25 and Annex One, p. 46.

[6] *Id.* at pp. 5, 50–51. The cross-subsidization issue will be discussed in connection with pricing policy in Ch. 9, *infra*. The interconnection issue will be discussed in Ch. 13, *infra*. Both have been dealt with in several countries, including the United States, through regulation.

[7] *Id.* at pp. 43, 50, 54–55.

[8] Mobile Green Paper, *supra* Ch. 3, note 25, at p. 40.

[9] Communication to the European Parliament and the Council on the Green Paper on Mobile and Personal Communications 6, 30, 45, COM(94) 492 final, Brussels, Nov. 23, 1994, [hereinafter "CEC Communication"].

[10] *EC Telecommunications Law*, Ch. 1, N. Higham, L. Gordon, S. White (eds.), Wiley Chancery Law 1993 & 1994 update.

[11] At the time of the decision, hearings for all applications in the top 30 markets had already been designated, and some initial decisions had been issued. In the Matter of Amendment of the Commission's Rules to Allow the Selection from Among Mutually Exclusive Competing Cellular Applications Using Random Selection or Lotteries Instead of Comparative Hearings, ¶¶ 12, 17, 21 & note 12, 56 R.R.2d 8 (1984).

[12] Notice of Proposed Rulemaking and Tentative Decision, In the Matter of Amendment of the Commission's Rules to Establish New Personal Communications Services, *supra* Ch. 6, note 25 at ¶¶ 84–90.

[13] *Id.* at Appendix D. Subsequent efforts to reduce the number of speculative applications, such as imposing fees on lottery entrants and limiting the resale of licenses, were not successful, according to the study. *Id.*

[14] The Changing Role of Government in an Era of Deregulation, *supra* note 1, at 85. The windfall arises out of the fact that the lottery winner, who pays nothing for the license, frequently sells its license, often to an adjoining large operator that wishes to be able to provide its customers with a large service territory. Thus, there is effectively an auction of spectrum, with the proceeds going to the lottery winner rather than the public coffers.

[15] Overview, *supra* Ch. 3, note 56, § 1, ¶ 11.

[16] CEC Communication, *supra* note 9, at 16.

[17] "Fast-growing Asia presents problems for would-be operators," *FinTech Mobile Communications*, July 28, 1994.

[18] Licensing and Declaration Procedures, *supra* Ch. 6, note 12, at p. 55. Although most countries have domestic bribery laws, only the United States prohibits its citizens from bribing foreign officials, under the Foreign Corrupt Practices Act.

[19] While nominally a beauty contest, the perception was that the process was more akin to ad hoc decision making.

[20] "South Korean Cellular Licensing Debacle to be continued Next Year," *Mobile Phone News*, Sept. 10, 1992; "South Korean cellular winner returns license," *FinTech Mobile Communications*, Sept. 10, 1992; "South Korea Postpones Award of Cellular Contract Until New President Takes Over," *Computergram Int'l*, Sept. 1, 1992.

[21] Jerri Stroud, "Bell Wins Stake in Korea Net," *St. Louis Post-Dispatch*, May 12, 1994, p. 8C; From news services and staff reports, *Wash. Post*, May 12, 1994, p. D9. Coincidentally, Sunkyong, which withdrew from the second round of tenders for the second license, was the successful bidder for a 23% stake in Korea Mobile Telecom (KMT) when the government privatized it in January 1994. Sunkyong was the only successful bidder out of 292 applications. The rest of the bids were too low, and it was reported that a new auction would be scheduled. John Burton, "Sunkyong takes 23% stake in Korean telephone network," *Fin'l Times*, Jan. 27, 1994.

[22] Steve Glain, "U.S. Partners Chafe as Seoul Demands Phone Venture Use Untried Technology," *Wall St. J.*, Sept. 28, 1995, p. A16; Steve Glain, "Consortium Says South Korean Rules Could Smother Its Cellular Venture," *Wall St. J.*, March 10, 1995, p. B6D.

[23] The Changing Role of Government in an Era of Deregulation, *supra* note 1, at 82.

[24] *Id.* at 85.

[25] "GTE Garners Argentine Nationwide Cellular License," *Global Telecom Rep't*, Nov. 15, 1993. Buenos Aires already had two cellular providers.

[26] Specification of Bases and Conditions for the International Public Bid for the Provision of Mobile Telephony Services in The Argentine Republic, title I, section no. 1, [hereinafter "Argentine Specifications"].

[27] It is questionable whether the government's expenses in regulating cellular increase in direct proportion to the revenues of the operators being regulated.

[28] Argentine Specifications, *supra* note 26, at title II, section no. 10; Memorandum to the File of Stephanie W. McCullough, United States Department of Commerce, June 24, 1993.

[29] Argentine Specifications, *supra* note 26, at title I, section no. 1. A mathematical equation was used to weight the factors and assign a final score.

[30] "GTE prepares to meet its 30-day Argentina challenge," *FinTech Mobile Communications*, Nov. 18, 1993; Peter Bate, "GTE, AT&T Plan Cellular "Desert Storm" in Argentina," *Reuter Asia-Pacific Business Rep't*, Nov. 8, 1993.

[31] Invitation to Apply for a License to Provide a National Cellular Radio Telephony Service issued by the South African Minister of Transport and of Posts and Telecommunications, Sections 43–57. A certain amount of redundancy can be detected in these criteria.

[32] This was designated a key determinant that would weigh heavily in the balance. See Ch. 10, *infra*.

[33] The Changing Role of Government in an Era of Deregulation, *supra* note 1, at 85.

[34] For a description of an "expedited" hearing procedure that the FCC had hoped would minimize delays and the criteria to be used in such proceedings, see Report and Order, In the Matter of An Inquiry Into the use of the Bands 825–845 MHz and 870–890 MHz, *supra* Ch. 3, note 53, at ¶ 42 & nn. 55–56, ¶¶ 63–78.

[35] Notice of Proposed Rulemaking, In the Matter of Amendment of the Commission's Rules to Establish New Personal Communications Services, Appendix D, GEN Docket No. 90-314, ET Docket No. 92–100 (1992).

[36] Licensing and Declaration Procedures, *supra* Ch. 6, note 12, at p. 54. The third standard listed, that criteria be based on maximizing customer benefit and choice, is less an argument for fair process than a substantive criterion that may argue against auctions.

[37] Overview, *supra* Ch. 3, note 56, at p. 8, § I, ¶ 8.

[38] CEC Communication, *supra* note 9, at 16.

[39] Licensing and Declaration Procedures, *supra* Ch. 6, note 12, at p. 55.

[40] Notice of Proposed Rulemaking and Tentative Decision, In the Matter of Amendment of the Commission's Rules to Establish New Personal Communications Services, *supra* note 25, Appendix D.

[41] *Id.*, Appendix E.

[42] Omnibus Budget Reconciliation Act of 1993, Pub. L. No. 103-66, § 6002, 107 Stat. 312 (1993).

[43] Fifth Report and Order and Second Report and Order, Implementation of Section 309(j) of the Communications Act—Competitive Bidding, PP Docket No. 93-253 (1994); Bulletin, Bell Atlantic Mobile, Nov. 2, 1994; Amanda Bennett, "Nobel in Economics is Awarded to Three For Pioneering Work in Game Theory," *Wall St. J.*, Oct. 12, 1994, p. B3.

[44] FCC Grants 99 Licenses for Broadband Personal Communications Services in Major Trading Areas, June 23, 1995 (unofficial announcement of Commission action). The second phase of licensing did not begin until December of 1995, due to legal challenges. The FCC had attempted to implement Congress's instructions regarding minorities and women by giving members of those groups special payment terms, setting aside a special block of spectrum for them and small businesses, and other advantages. However, in light of a Supreme Court decision striking down as unconstitutional other preferences in federal construction contracting, the FCC revised its rules. Sixth Report and Order, In the Matter of Implementation of Section 309(j) of the Communications Act - Competitive Bidding, PP Docket No. 93–253, Amendment of the Commission's Cellular PCS Cross-Ownership Rule, GEN Docket No. 90–314, and Implementation of Sections 3(n) and 332 of the Communications Act Regulatory Treatment of Mobile Services (1995). The FCC rescheduled the second round of auctions to begin August 29, 1995. However, further challenges to the FCC's rules resulted in delays in the auction process. Daniel Pearl, "Federal Court Ruling Will Complicate Auctioning of PCS Licenses by the FCC," *Wall St. J.*, Nov. 13, 1995, p. B12; "Court Clears FCC to Plan Wireless Services Auction," *Wall St. J.*, Sept. 29, 1995; Daniel Pearl and Gautam Naik, "FCC's Wireless Communication Auction Hits Snag as Federal Court Intervenes," *Wall St. J.*, July 28, 1995, p. A16, col. 1.

[45] Licensing and Declaration Procedures, *supra* Ch. 6, note 12, at pp. 55, 58, 63.

[46] *Id.* at p. 66. Even when auctions are used, however, social obligations and specified tariffs can be imposed as license conditions.

[47] A word of clarification is appropriate here. In the United States, the word "price" is used to describe the charge for the service and the word "tariff" is used to describe the document filed with a regulatory body that sets out those charges. In Europe the word "tariff" is used synonymously with the word "price."

[48] Licensing and Declaration Procedures, *supra* Ch. 6, note 12, at p. 65.

[49] Overview, *supra* Ch. 3, note 56, at p. 8, § I, ¶ 10.

[50] *Id.* at ¶ 11.

[51] The Changing Role of Government in an Era of Deregulation, *supra* note 1, at 86.

[52] Jordanian tender, *supra* Ch. 6, note 17, at § 7.3. *See* Alan Stewart and Alan Pearce, "PCS: first, find your market," *Communications Int'l*, Oct. 1994, p. 78 (qualifications to bid on Israeli license included requirement that bidder's operational partner must have provided service to at least 100,000 subscribers as of day of bid submission and that bidder's net worth must be at least $200 million).

[53] Jordanian tender, *supra* Ch. 6, note 17, at § 7.4. The same approach—a beauty contest followed by an auction—appears to have been used in Hungary. United States Department of Commerce, Summary of Recent International Cellular Developments, Updated Nov. 18, 1993.

[54] Articles 14–17.

CHAPTER 8

▼▼▼

LICENSE PAYMENTS

If licenses are to be sold rather than given away, there are virtually unlimited ways to structure payment by the licensee to the government. Although at least one country has declined to impose a large fee at the time of the award of a cellular license [1], the more typical arrangements involve a significant up-front fee and a percentage share of either revenues or profits over time. (An annual flat fee may be substituted for a percentage share.) A frequency fee or subscriber fee may be imposed in addition to or instead of a share of revenues or profits. Sometimes the ongoing payments (or a portion of them) are characterized as an administrative fee to reimburse the government for costs incurred in regulating cellular. In addition to up-front and ongoing payments, in some countries the operator must also provide cellular service to the government for its own use at rates that are lower than those normally charged to the public. If the landline telephone company is state-owned, the government may try to use interconnection and infrastructure fees paid by the cellular operator as a way to obtain a share of the cellular operator's revenue rather than charging cost-based rates. Some countries also charge all potential bidders a fee merely to obtain the invitation to tender [2] and another fee to all actual bidders for the privilege of bidding [3].

How the actual license payment is determined depends on whether a beauty contest is used by itself, whether an auction follows a beauty contest, or whether an auction is used by itself. It also depends on what form of auction is used, sealed bids or oral or electronic auctions.

Where the sole basis of selection is a beauty contest, the government simply specifies the fees it thinks are appropriate and chooses among competing bids based on other factors [4]. Such fees are likely to be imposed even when the license is not awarded on a comparative basis [5]. The sorts of fees that it is likely to impose may include: an up-front payment when the license is awarded; a percentage share of either revenues or profits or a flat fee, due annually for the term of the license; a frequency fee; or a subscriber fee. Ongoing fees are sometimes labeled "administrative," to reimburse the government for costs incurred in regulating cellular [6].

In the case of auctions, there is somewhat more flexibility in how to structure the actual license payment where sealed bidding is used than when an oral or electronic auction is used. In an oral or electronic auction, there is not enough time to evaluate the value of the various components of the bid to determine which bid is actually the highest. Where sealed bidding is used, the government often specifies certain minimum payments that the licensee will be required to make and then allows the licensee to offer more than the minimum in order to win the license. These payments are generally of the same nature as those described above as being used in a beauty contest.

A government that chooses to stipulate minimum or actual fees for licensees is faced with a plethora of policy decisions. It must make a tradeoff between current and future revenue, for one. If it demands a higher payment when the license is awarded because it has a short-term need for cash, it should be prepared to impose lower payment requirements in future years. Various combinations of payments can yield the same net present value (NPV). But the government should also be alert to other pressures. Private investors may be compelled by their shareholders to aim for short-term profits, so that they may need to recover their initial investment quickly. This means that they may be less interested in bidding on a license that requires a large up-front payment than in one that requires larger payments over time, even though the NPV of each payment stream is the same [7] (or the NPV of the one requiring the larger up-front payment is smaller). However, the government may feel that a large up-front payment will incent the operator to build out its system more quickly in order to obtain high service revenues as quickly as possible [8]. The government must also decide whether a flat annual fee or a percentage of the cellular operator's revenues or profits every year will be more advantageous. Does the government have sufficient confidence in the operator that it is willing to share the risk that it will be successful [9]?

Three final ways of imposing additional costs on cellular licensees that result in benefit to the government are worthy of mention. These advantages can be gained by the government regardless of the method by which the license is awarded. They are interconnection fees, infrastructure fees, and discounted service.

In cases where the landline telephone company is state-owned, the government may try to use interconnection fees paid by the cellular operator as a way to obtain a percentage of the cellular carrier's revenues or profits rather than charging cost-based interconnection rates [10]. As discussed in Chapter 13, interconnection with the landline network is a prerequisite for mobile-to-land and land-to-mobile calls.

The government may also require cellular carriers to lease the lines they use to connect the cell sites and switches of their cellular networks with each other from the local or national telephone company. In the United States, cellular carriers are generally free to build infrastructure links themselves or to obtain them from the telephone company, a wireline long-distance company, a microwave company, or a satellite operator [11]. But in other countries the cellular operator may be legally required to obtain such links from the local or national telephone company, which may be owned by the government [12].

Non-cost-based interconnection fees and mandatory infrastructure purchases raise additional policy issues. Is it advisable to allow the landline telephone company to be subsidized through interconnection rates that are unrelated to its costs and to have a monopoly on cellular network infrastructure? Perhaps the government will feel obligated to compensate the landline carrier for the customers it will lose to a cellular operator that provides competitive service, but is this the best use of those cellular revenues? The EU's Mobile Green Paper concluded that cellular carriers in member states should have free choice of infrastructure facilities [13]. The CEC, after public consultation on the Green Paper, stated that it considers free choice of infrastructure "as essential for ensuring an appropriate economic base" for mobile network operators [14].

The telecommunications ministers of the EU, however, failed to agree to deregulate infrastructure for mobile or personal communications ahead of the January 1, 1998, deadline for opening of public telephone services and all infrastructure up to full competition. Rather, they concluded that member states would make their best efforts to follow the CEC's guidelines and deregulate infrastructure for mobile operators by ending monopolies and allowing operators to build their own infrastructure and interconnect with each other prior to 1998. Six member states (the Netherlands, Finland, the United Kingdom, France, Italy, and Sweden) issued their own resolution in support of an earlier deadline of January 1, 1996 [15].

Faced with this position by the Ministers, the CEC exercised its Article 90 powers to order member states to allow mobile operators to use all available means of transmission for infrastructure and interconnection. The directive took effect in early 1996. Member states would then have nine months in which to inform the Commission of measures taken to comply, although Greece, Ireland, Portugal, and Spain have been given up to five years in which to report [16]. (The CEC's Article 90 powers are discussed in Section 7.2, *supra*.)

The government may also impose an obligation on cellular operators to provide service to the government at special, discounted rates. The concession granted by the Mexican government to Iusacell's predecessor provides that the license shall be revoked if the company does not provide the government a discount of 50% off the regular fees charged to the public [17]. The United States permits, rather than requires, its cellular carriers to provide federal, state, and local governments with favorable rates. Telecommunications law in the United States prohibits common carriers, a category that includes cellular licensees, from making "any unjust or unreasonable discrimination in charges," giving "any undue or unreasonable preference or advantage to any particular persons, or class of persons," or subjecting "any particular person [or] class of persons...to any undue or unreasonable prejudice or disadvantage" [18]. However, the Communications Act carves out an exception to this rule for service to the "government" [19]. A cellular operator may offer special rates to the government without violating the prohibition on discrimination. These sorts of provisions raise another policy question for the government: whether it is preferable to require or allow cellular carriers to offer discounted service to the government or to require complete nondiscrimination and simply subsidize government purchases of cellular service through revenues collected from cellular carriers (or other sources) [20].

Whatever the fees and obligations imposed on a cellular service provider, it is important that they be imposed equally on all service providers. The European Union is implementing this policy by insisting that the governments of Italy and Belgium charge the cellular licensees that are affiliated with their respective state telephone companies the same fees that they are demanding from private cellular licensees [21].

Notes

[1] Argentina, as noted *supra* Ch. 7.

[2] While these fees are not insignificant (usually at least a few thousand dollars), nor are they generally large enough to discourage any serious bidder.

[3] These may ultimately be considered a down payment on a winning bid. This was the case with fees imposed on registrants for the broadband PCS auctions. "FCC revealed size of up-front payments by each potential bidder in broadband PCS auction," *Common Carrier Week*, Dec. 5, 1994.

[4] Licensing and Declaration Procedures, *supra* Ch. 6, note 12, at Annex One, p. 14 (Denmark), pp. 21–23 (France), pp. 36, 39 (Germany), pp. 50–52 (Greece), p. 82 (the Netherlands), p. 89 (Portugal), and p. 107 (the United Kingdom).

[5] Article Sixteenth of the 1957 concession granted to Servicio Organizado Secretarial, S.A., as amended July 11, 1986.

[6] Note 4, *supra*. See also FCC Public Notice: FY 1995 Commercial Wireless Regulatory Fees, Aug. 1, 1995 (setting forth regulatory fees to be collected pursuant to statute to offset FCC's costs); United States Department of Commerce, Summary of Recent International Cellular Developments, Updated Nov. 18, 1993, at p. 11 (discussing fees to be imposed by Israeli government). These fees have historically been less likely to be imposed when the licensee is also the landline telephone company. Licensing and Declaration Procedures, *supra* Ch. 6, note 12 at Annex One, p. 6 (Belgium) and p. 80 (the Netherlands).

[7] Licensing and Declaration Procedures, *supra* Ch. 6, note 12, at pp. 28–30.

[8] This concern can also be addressed through the use of geographic coverage requirements that must be met to retain the license. Ch. 10, *infra*.

[9] Turkey seems to have solved this dilemma by imposing a large up-front fee of $500 million dollars and a requirement that the licensee pay the government a percentage of revenues if the number of subscribers exceeds 150,000. United States Department of Commerce, Summary of Recent Inter-

national Cellular Developments, Updated Nov. 8, 1993, at p. 19. Not every country will be able to be that demanding.

[10] Exhibit F to Joint Venture Agreement Between Post and Telecommunications Administration of the Slovak Republic and Atlantic West B.V. (fee to interconnect with landline system "will remain constant at 3% of the airtime usage revenue generated by the cellular system"); Licensing and Declaration Procedures, *supra* Ch. 6, note 12, at Annex One, p. 52 (interconnection fee of 5% of cellular airtime revenue in Greece).

[11] The cellular affiliates of regional Bell operating companies have had special restrictions on their abilities to carry calls across LATA boundaries and to hire another company to do so. These restrictions were loosened by the 1996 Telecommunications Reform Act.

[12] Licensing and Declaration Procedures, *supra* Ch. 6, note 12, at p. 5.

[13] Overview, *supra* Ch. 3, note 56, at p. 11, § IV, ¶¶ 1, 3.

[14] CEC Communication, *supra* Ch. 7, note 9, at 35.

[15] "EU Ministers Fail to Back Call to Speed Up Deregulation Efforts," *BNA Int'l Trade Reporter*, June 14, 1995, p. 1028.

[16] "EU Mobile Phone Measure," *Wall St. J.*, Jan. 17, 1996, p. A11; "Mobile telephones must compete from February: EU," *AFP*, Dec. 20, 1995.

[17] Article Twenty-Seventh, 1957 Concession granted to Servicio Organizado Secretarial, S.A.

[18] 47 U.S.C. § 202(a).

[19] 47 U.S.C. § 202(b).

[20] One might also abolish nondiscrimination requirements with respect to cellular companies on the theory that they should not be considered common carriers.

[21] "Belgacom chief angry over fee," *Fin'l Times*, Oct. 10, 1995; "Brussels acts against Italy on mobile phone competition," *Fin'l Times*, Oct. 5, 1995; "Belgacom to pay mobile phone fee," *Fin'l Times*, Sept. 11, 1995, p. 2.

CHAPTER 9
▼▼▼

RATE REGULATION

9.1 INTRODUCTION

Should the government regulate the rates that cellular licensees may charge for service and equipment? If so, what method should it use? Should it set cellular service rates using the same methodology it does for basic landline service, or in some other fashion? Should cellular operators be able to charge more for service when the customer does not purchase its cellular phone from the operator? What if any special rules should apply when the cellular operator is also the landline service provider?

Two mechanisms for making regulatory conditions regarding rates, as well as other terms and conditions of the license, legally enforceable are available. They have been described as follows [1]:

1. The Anglo-American approach, under which general authorizations or franchises are granted subject to an overriding set of regulations which are changed to meet new circumstances; and
2. The Latin or civil-law approach under which a carrier is granted a franchise for a fixed period with detailed quantitative requirements that are to be maintained, but not fundamentally altered, by the regulator. The first approach often leads to more active regulatory involvement; the second approach, although easier to administer, may, due to technical innovations...leave the regulator with an outmoded instrument to enforce the government's policy choices.

With this background in mind, the questions set forth above can be addressed.

9.2   SERVICE RATES

The answers to the questions about regulation of rates for cellular service [2] would seem to depend in large part on whether cellular is expected to serve as a substitute for landline service and whether there is effective competition in the relevant market. That is, if cellular is seen as a luxury good, one might expect little or no price regulation. If cellular is seen as an essential service, one might anticipate the presence of price regulation, with the degree of regulation dependent on whether the discipline of competition is present [3]. In practice, however, the correlations are not nearly so neat.

Before proceeding to examine the examples, it may be helpful to review the regulatory options that are available. Government may simply allow the operator to establish and change rates freely, without any need for approval by or notice to the telecommunications regulatory agency. It may allow rates to be set as the operator sees fit, but require that informational tariffs be filed with the regulatory agency. It may informally negotiate rates and changes in them with the operator. It may set a ceiling and floor and allow the operator to charge rates within that range without approval and with or without notice to the agency. (This is sometimes referred to as *min-max* regulation.) Or it may use rate of return or price cap regulation, borrowing from principles used to set prices for landline service [4].

If cellular is not considered an essential good, then one might take the position that price regulation of any sort is not appropriate [5]. In fact, in countries where government has licensed cellular without the intent that it serve as a viable alternative to landline service for a significant part of the population because there was no need for such an alternative—high teledensity countries—regulation of rates has run the gamut. It has variously been nonexistent, informal, moderate, or restrictive, depending on the jurisdiction.

The United States serves as a good example of some of the divergent practices regarding regulation of cellular pricing where cellular is not a substitute for landline service. Because of the country's high landline telephone penetration, cellular has not been considered an essential good in the United States [6].

When the FCC originally authorized cellular service on a duopoly basis—two providers in every market—it permitted the 50 states and the District of Columbia to continue to regulate intrastate provision of cellular service. Over the years, the number of states that declined to exercise jurisdiction over cellular or that discontinued regulation after having it in place for a period of time steadily increased. By 1993, fewer than half the states regulated cellular. Typical reasons given by state public utilities commissions for deregulating cellular were that two providers are sufficient to make a competitive market, that the filing of tariffs is anticompetitive because it burdens pricing flexibility and gives the competitor advance notice of pricing, that cellular is not a necessity, and that cellular service rates are generally lower in states where they are not regulated. Jurisdiction to hear complaints by members of the public against cellular service providers was often retained by the commissions, as was power to regulate interconnection of cellular systems with the landline network [7]. Those that did regulate rates tended to use min-max regulation [8] and sometimes regulated only wholesale rates [9]. However, none of them used the sort of complex price cap or rate of return regulation that they applied to landline telephone companies. In a 1992 proceeding, the FCC determined that the two carriers in the duopoly markets do compete with each other "on the basis of market share, technology, service offerings, and service price" but that "[t]he record is not conclusive as to whether the service market is *fully* competitive" [10].

In the summer of 1993, a federal law preempting much of the remaining state regulation of cellular (which was now classified as a "commercial mobile radio service") was enacted. The federal Omnibus Budget Reconciliation Act [11] (the Budget Act) essentially divided state regulation of cellular into three categories: entry; rates; and terms and conditions. States were allowed to continue regulation of terms and conditions. States were preempted, effective August 10, 1994, from regulating entry, leaving that task to the FCC. States were preempted, effective August 10, 1994, from regulating rates, although they could petition the FCC for permission to continue rate regulation. The FCC does not set the cellular rates that a carrier may charge, although it does prohibit unreasonable and unreasonably discriminatory rates. Petitions to continue rate regulation were required to be filed by August 9, 1994. The Budget Act required the FCC to grant such permission if the petitioning state could demonstrate the following [12]:

(i) Market conditions with respect to such [commercial mobile radio services] fail to protect subscribers adequately from unjust and unreasonable rates or rates that are unjustly or unreasonably discriminatory; or

(ii) Such market conditions exist and such service is a replacement for land line telephone exchange service for a substantial portion of the telephone land line exchange service within such State.

Thus, even though cellular is not a substitute for landline service in the United States, regulation of rates by state commissions is still permissible under subsection (i) where the competition that would bring about just, reasonable, and nondiscriminatory rates is not believed to be present. In addition to allowing states that regulated cellular at the time the 1993 Budget Act was passed to seek continuation of such authority, the Budget Act also permits states that did not previously regulate cellular to petition to be allowed to do so [13].

In its order elaborating on how a petitioning state could succeed in its request to be allowed to continue regulation, the FCC said that while states would have discretion to submit whatever evidence of market conditions they feel would be persuasive

[14], we would consider the following types of evidence, information, and analysis to be pertinent to our examination of market conditions and consumer protection:

1. The number of CMRS [commercial mobile radio service] providers in the state, the types of services offered by these providers, and the period of time during which these providers have offered service in the state.
2. The number of customers of each such provider, and trends in each provider's customer base during the most recent annual period (or other reasonable period if annual data is not available), and annual revenue and rates of return for each such provider.
3. Rate information for each CMRS provider, including trends in each provider's rates during the most recent annual period (or other reasonable period if annual data is not available).
4. An assessment of the extent to which services offered by CMRS providers that the state proposes to regulate are substitutable for services offered by other carriers in the state.
5. Opportunities for new entrants that could offer competing services, and an analysis of existing barriers to such entry.
6. Specific allegations of fact (supported by an affidavit of a person or persons with personal knowledge) regarding anti-competitive or discriminatory practices or behavior on the part of CMRS providers in the state.
7. Evidence, information, and analysis demonstrating with particularity instances of systematic unjust and unreasonable rates, or rates that are unjustly or unreasonably discriminatory, imposed upon CMRS subscribers. Such evidence should include an examination of the relationship between rates and costs. We will consider especially probative the demonstration of a pattern of such rates, if it also is demonstrated that there is a basis for concluding that such a pattern signifies the inability of the CMRS marketplace in the state to produce reasonable rates through competitive forces.
8. Information regarding customer satisfaction or dissatisfaction with services offered by CMRS providers, including statistics and other information regarding complaints filed with the state regulatory commission [14].

Eight states (Arizona, Connecticut, New York, Hawaii, Louisiana, Ohio, California, and Wyoming) filed petitions. Wyoming withdrew its petition before the FCC could rule on it [15].

The FCC denied each of the state petitions to continue rate regulation. It began by defining what would constitute unjust and unreasonable rates for purposes of the Budget Act. Rates would be characterized as just and reasonable if they fell within a "'zone of reasonableness' that is bounded at one end by the 'investor interest in maintaining financial integrity and access to capital markets' and at the other by the 'consumer interest in being charged non-exploitative rates'" [16]. Reasonableness is not "dictated by reference to carriers' costs and earnings, [footnote omitted] but may take account of non-cost considerations such as whether rates further the public interest by tending to increase the supply of the item being produced and sold" [17]. The FCC was also heavily influenced by the fact that PCS licensees would be providing service within the next two years. It felt that the impending competition represented by PCS weighed strongly against the states' requests to be allowed to continue regulation, absent any indication that demand would be so low that entry by PCS licensees would be unlikely. The FCC also noted that it would be influenced by any evidence of collusion between the two existing cellular operators. In general, it stated, it would find dynamic evidence (such as growth and investment) more persuasive than static evidence (such as rates of return) given the rapidly changing nature of the market. Finally, it added in the balance the Congressional policy in favor of deregulation and a national regulatory policy for CMRS. After considering these factors, the FCC determined that none of the petitioning states had met the burden of proving the necessity of continued rate regulation [18].

As was the case among the various United States prior to the FCC's recent preemption decisions, a disparity of approaches is also apparent when one compares other high-teledensity countries. In Belgium, for instance, where the state telephone company has had a monopoly on cellular service, cellular rates (along with other public telecommunications services) are based on a formula that tracks consumer prices. In France, the analog and digital cellular operators are free to set their prices. The private GSM operator in Germany must publish its prices and inform the regulator before they enter into effect. Denmark requires approval of prices by the Minister of Posts and Telecommunications before they enter into force, and the Minister may request changes in them. Italian tariffs are set by Ministerial decree [19]. In New Zealand, cellular rates (along with the prices of other telecommunications services) have been left to the marketplace [20].

In other countries, cellular serves as a substitute for landline service because of the inability of the landline telephone company to satisfy demand. The government may or may not have intended that cellular fulfill this need when it licensed cellular. It may have authorized cellular licensing as a way to solve landline's deficiencies, or cellular may have become the solution by default. Regardless, cellular may be considered an essential or basic service in these countries.

An example of a country in which the demand for basic telecommunications service has not been met and in which one might therefore expect cellular to fill the gap is Mexico. For many years the regulation of cellular rates in Mexico did not evidence an intent on the part of the government to facilitate that result, either by keeping rates low or by allowing competition to drive rates down. Until 1995, the Ministry of Communications and Transportation (SCT) set rates for cellular airtime, so that each of the two carriers in any given region charged the same rates [21].

The process for setting rates was not transparent, so it was not possible to determine the exact methodology used by the SCT. The SCT's stated objectives in rate regulation included the promotion of a viable and efficient industry capable of providing consistent and affordable service [22]. The Mexican Telecommunications Regulations stated that approval of rates would be governed by the following bases, which in part seemed to be more aspirational than operational [23]:

1. That tariffs are oriented at cost and that no discounts or bonuses will apply that imply collections below direct cost.
2. To avoid the existence of cross subsidies between services provided by the same concessionaire, except when expressly authorized by the Ministry.
3. The international competitiveness of the tariff levels.
4. The generation of reasonable margins of revenue, in accordance with prevailing economic conditions.
5. To promote a sufficient expansion of the services and provide the basis for healthy competition in the rendering of the same.
6. Tariffs should not be discriminatory nor hinder access to the service in question.
7. The methods for calculating the tariffs should be clearly stipulated.

Historically, upon request of a cellular operator, the SCT permitted rate increases in that operator's region, taking into account such factors as the level of competition in the region, the financial condition of the operator, and certain other economic considerations such as the rate of devaluation of the peso against the U.S. dollar [24].

In fact, the extent to which cellular has substituted for landline service in Mexico has been limited. This is demonstrated by the cellular penetration rate, which was less than 1% in mid-1993 [25]. To some degree, this may be attributable to the fact that rates are higher than the average Mexican can afford [26].

In 1995, however, the Telecommunications Law of Mexico was revised to allow license holders the freedom to set their own service rates, subject to prior registration of those rates with the government [27]. It is too early to determine whether this will result in lower rates and higher cellular penetration, although it will be an interesting test.

When a country introduces cellular service with the intent that it fill the need for basic service, but does not believe that the market can support two cellular competitors, is traditional rate of return or its newer alternative, price cap regulation, appropriate? Theoretically, the argument for this approach can be made. But finding an example of where such regulation has been implemented is difficult.

There are any number of possible reasons why governments have not chosen to use rate of return or price cap regulation for cellular service rates. One such explanation is that cellular costs have been for some time too high to allow the technology to be a substitute for landline service. Cellular has been perceived as a luxury for a long time, in part due to its costs. Although costs have been reduced, the perception lingers. Another possible reason is that rate of return and price cap regulation are means to regulate the rates of a monopoly and cellular may not be considered a monopoly in this circumstance. That is, even if there is only one cellular provider, it does not have a monopoly on basic service because it must compete with the landline provider in that market. Its monopoly is only in the mobile market, and mobile service is not essential. To put it another way, to the extent that a sole cellular operator provides essential service, it does not have a monopoly; to the extent that it has a monopoly, it is not providing an essential service. This rationale would not apply, however, in countries where the landline provider is an ineffective competitor. Two additional possible explanations for the dearth of countries using rate of return or price cap regulation for cellular are: (1) the administrative burdens are too great, for both the operator and the government; and (2) countries may decide that cellular rates only need to be regulated for customers who can not afford to pay rates based on the operator's cost, and these customers are protected by universal access requirements [28].

One common practice today is to request prospective cellular licensees to specify in their bids the rates they will charge and the amount by which they will limit rate increases over the next several years. These terms are then written into the license. This was, for example, the approach taken in the Jordanian invitation for tenders that the IFC (World Bank) assisted in drafting. That document invited bidders to specify initial rates and expected inflation, stating that bidders must commit to keep rates unchanged (except for increases to reflect inflation and increases approved by the Ministry on a case-by-case basis) for four years. Rates could be lowered without permission from the Ministry [29]. This approach would not be feasible when an auction or lottery is used to award the license. However, it would be a suitable criterion for use in a comparative process, such as a beauty contest. Indeed, this was the approach recommended by the study commissioned by the European Union [30]. While the solution may lack a degree of flexibility, it does reduce the administrative burden associated with rate regulation.

In sum, there seems to be no consistent practice with respect to regulating cellular service rates. Whether cellular service is a necessity seems to be taken into account by some regulators, but is irrelevant to others. It rarely seems to be dispositive of the question of whether regulation is needed, but is simply one factor in the mix. The existence of competition is also sometimes considered pertinent, but not on a regular basis. Often, the license is simply awarded on the basis of which prospective operator promises the lowest rates for the longest time.

One final note on service pricing should be addressed. When a call is made from a landline phone, the calling party rather than the called party is the one billed, unless the charges are reversed. In cellular, the presumption is usually that the cellular subscriber pays airtime charges regardless of whether she initiated or received the call. This probably arose because cellular companies could bill only their own customers and not the calling parties. However, a service referred to as *calling party pays* (CPP) has been developed by which the landline telephone company can bill the party calling the cellular subscriber and remit the proceeds (less a fee) to the cellular company [31]. Evidence indicates that where the CPP principle is applied, outgoing subscriber minutes increase and subscribers are more willing to have their cellular number listed in a phone book [32]. CPP appeals to cellular subscribers because they feel they can control their bills and still be accessible. Developing countries may want to consider requiring licensees to provide service on a CPP basis, as a way to encourage higher teledensity [33].

9.3 EQUIPMENT RATES

With respect to pricing of cellular equipment, the general trend has been not to regulate it. Where the prices of phones have been regulated, the result—intentional or inadvertent—has in some instances been to maintain prices at artificially high levels. This makes it more difficult for people to procure cellular service. The exception to this has been where the cellular operator has tried to recoup its investment in the system through very high handset prices. In that case, regulation might succeed in bringing handset prices down, although competition should also accomplish the same result.

If the theory behind regulating prices for cellular service is that those prices are not subject to the discipline of a competitive market, this theory does not justify regulating prices for cellular phones. Cellular phones for the commonly used technologies are produced by a number of manufacturers and are distributed through several channels: the cellular operator, its agents, retail stores (either department stores or specialty electronics store), and catalogs. They can be purchased (sometimes on an installment basis) or leased. Given the large number of equipment providers in the market, prices are subject to competitive discipline [34].

Ironically, regulation of cellular phone prices has sometimes put phones out of the reach of most people. For example, Japan for many years permitted only the leasing, rather than the sale, of cellular phones. For some time Japan experienced such an economic boom that cellular sales grew despite the cost. However, when the economy slowed in the early 1990s, the rate of cellular expansion fell off. The Ministry of Posts and Telecommunications agreed in 1992 that as of April, 1994, it would allow cellular phones to be purchased. This has prompted another boom in cellular sales in Japan [35].

The opposite problem occurred in Indonesia. The first two cellular licenses issued in that country were to investor groups owned in whole or part by children of the president of the country. These groups built their cellular systems and then turned them over to the state-owned telephone company to run. The investor groups were given a share of the cellular service revenues from the project and were allowed to keep the revenues from activation fees and the sale of cellular phones. The result was that the price of phones and the activation fees were quite high, dampening cellular growth [36]. Indonesia is said to have the lowest penetration of cellular phones in Asia [37]. Recently, the government has placed a ceiling on the price of handsets [38].

9.4 PACKAGING OF CELLULAR SERVICE AND EQUIPMENT

On occasion, a customer who already has a cellular phone wants to purchase only cellular service: perhaps he or she subscribed to the competitor's service and was unhappy with it. Sometimes, a customer who already has service wishes to purchase only a new phone: to upgrade to a more compact unit or one with new features or to replace a stolen or damaged phone. But in most cases, customers both purchase a phone and subscribe to service with a new phone number. Accordingly, cellular companies usually offer their customers packages comprising a phone and a service contract. This raises two pricing issues: tying and bundling.

In United States antitrust law, *tying* refers to the situation where a seller with market power in one market (sometimes, but not always, a monopoly) requires a customer to purchase a good for which the seller does not have market power (the *tied* product), before it will sell the customer the one in which it does (the *tying* product) [39]. In the context of cellular, a tying violation would occur if a monopoly or possibly a duopoly cellular operator required a customer to purchase a phone in order to get service [40]. The competitive harm that arises from a tying arrangement is that it limits the purchaser's freedom to buy a product (in this case, a cellular phone) from the source of his or her choice. Another source may offer a better product or a lower price. Thus, tying is unlawful because it is a restraint of trade [41].

The United States FCC also prohibits the practice of requiring a customer to purchase a phone in order to obtain cellular service. Cellular service must be available on an "unbundled," nondiscriminatory basis. As under antitrust law, the FCC justifies the policy on the basis of customer choice. However, it also applies a broader public interest standard than the promotion of competition. The FCC does not prohibit cellular operators from offering packages of equipment and service. It simply requires that the customer be able to buy service or equipment separately if he or she so desires. It also mandates that the operator charge the same price for service regardless of whether the customer also purchases equipment.

However, the converse is not prohibited. That is, a cellular operator may charge less for the equipment if the customer agrees to subscribe to cellular service.

The public interest enunciated by the FCC in support of this rule was that subsidization of phone prices with revenue from the sale of service would allow people who could not otherwise afford cellular to subscribe to it. This would result in a more efficient use of cellular spectrum and would encourage universal availability and affordability of cellular [42]. The existence of this public interest depends on cellular operators being able to price their service at high enough rates that they can afford to subsidize the equipment, a condition that may not be present in a developing country.

In fact, this practice is quite common in the United States and other countries. Carriers are willing to sell cellular phones below their cost and subsidize the loss with profits on the service that they sell [43]. Agents who represent the carriers adopt a similar practice: they subsidize their loss on the phone with the commission that the carrier pays them for activating a new customer's service. Some complaints have been registered that this subsidy of the equipment price comes as a surprise to customers whose phones have been stolen. Because they are not activating a new service account, they must pay full price for their replacement phone, in comparison to the discounted price they paid for the original one [44].

A policy on bundling that resembles the one espoused by the FCC has recently been adopted in Hong Kong. The FCC, as noted, requires cellular operators to charge a customer the same price for service even if the customer buys a phone from another source. In Hong Kong, cellular operators have charged customers much higher service activation fees to connect phones purchased from outside suppliers than to connect phones purchased from the cellular operator. The reasons offered by the operators were that it cost them more to program outside-supplied phones for the network, to provide them lines, and to handle their administration. Hong Kong's Office of the Telecommunications Authority did not accept these explanations, and pressured the colony's four cellular operators to drop activation prices for customers who use phones from another vendor [45].

9.5 CELLULAR SEPARATION RULES

Where the landline telephone company also holds a cellular license, it may be required to operate its business as a separate subsidiary. This separate subsidiary requirement, mandated by what may be referred to as *cellular separation* rules, can prevent the packaging of landline and cellular services that could be offered to the public in combination at a rate lower than would be paid if each were purchased individually. That is, the carrier is prevented from taking advantage of economies of scope. Indeed, this handicap applies not only to joint marketing, but to sharing of resources for many other purposes as well. If the result is higher prices for the public, why should such a requirement be imposed? The answer is that the rule is imposed in the hope that it will prevent unfair competitive practices.

The cellular separation rule promulgated by the United States FCC is imposed only on the regional Bell holding companies [46]. It prohibits these companies from engaging in the provision of cellular service except by way of a "separate corporate entity" and further states, in relevant part, as follows [47]:

(2) Each such separate corporation shall operate independently in the furnishing of cellular service...Each such separate corporation shall maintain its own books of account, have separate officers, utilize separate operating, marketing, installation, and maintenance personnel, and utilize separate computer transmission facilities in the provision of cellular services. Any research or development performed on a joint or separate basis for the subsidiary must be done on a compensatory basis;...
(d) A carrier subject to the restriction...above:
(1) Shall not engage in the sale or promotion of cellular services on behalf of the separate corporation or sell, lease or otherwise make available to the separate corporation any transmission facilities which are used in any way for the provision of its landline telephone services, except on a compensatory, arm's length basis; this section shall not prohibit joint advertising or promotional efforts by the landline carrier and its cellular affiliate...

The underlying purpose of the cellular separation rules is to prevent the Bells' landline companies from using revenues from their monopoly, rate-regulated businesses to subsidize their competitive cellular businesses. It was feared that such subsidization would give cellular affiliates of landline companies an unfair advantage. To enforce this provision, Bell wireline companies were required to reduce to writing and file with the FCC copies of all contracts with their cellular affiliates. The filing requirement was later changed to require the filing only of interconnection agreements [48]. The FCC also expressed the concern that companies would misallocate costs: that they would take costs properly allocable to cellular service and assign them to rate-regulated landline operations, resulting in unfair burdens on basic service customers. It recognized that "[a] cost of this separation is the preclusion of some of the possible economies in jointly providing both traditional wireline and cellular service on an integrated basis." However, it concluded that "[b]ecause these services are not now jointly provided...the cost savings of foregone joint production may not be large, i.e., stand-alone provision of these services is already technologically feasible and existing wireline plant has not been engineered with the joint supply of cellular service in mind" [49].

The rationale behind the cellular separation rules has been undercut somewhat by the fact that the FCC did not require the Bell companies to hold their PCS businesses in separate subsidiaries. One Bell company has sought elimination of the cellular separation rules on this basis. Although the FCC declined to abolish the rules, a federal appeals court found the difference in the FCC's approaches for cellular and PCS to be unwarranted and ordered the FCC to commence an inquiry as to whether the cellular rule should be repealed [50].

The study commissioned by the European Union precedent to the issuance of its Mobile Green Paper also addressed the merits of cellular separation. It identified numerous operational and marketing advantages to vertical integration of landline and cellular businesses. These were the ability to offer customers package deals, sharing of staff, collocation of equipment enabling cost savings from shared maintenance, sharing of site acquisition expenses, sharing of billing processes, and use of customer databases. However, it also felt that such cross-subsidization causes anticompetitive distortions. It recommended "open, transparent, and arms-length relationships between TOs [the incumbent national telecommunications organizations] and affiliated mobile operators" [51]. Nonetheless, the Mobile Green Paper "limited itself to fields where a common position is required at a European Community level" [52]. Accordingly, it applied the policy on such relationships to interconnection and not to broader issues of cellular separation.

In Mexico, one of the nonwireline cellular carriers, Iusacell, has sued Telcel, the cellular affiliate of the landline carrier, Telmex, alleging that Telmex utilizes it monopolistic power in the local telephony market to subsidize and grant other advantages to Telcel, consequently discriminating against Iusacell and other competitors of Telcel. The lawsuit was filed in November 1995, with the Mexican Federal Competition Commission [53].

Notes

[1] Smith and Staple, *supra* Ch. 1, note 7, at 67.

[2] Rates for cellular service are typically composed of an initial activation fee, a monthly access charge, and a per minute airtime charge. There are generally two airtime rates, one for peak and one for off-peak times.

[3] Excluded from this discussion is whether the government should set rates for customers for whom cellular is an essential service, but who cannot afford to pay for cellular service that is priced at rates that reflect the cellular operator's cost plus a fair return. These customers are addressed in the next chapter.

[4] For a description of price cap and rate of return regulation, see Ch. 5, *supra*. Because rate of return regulation is so rarely used in the cellular field, there is no agreed upon methodology for calculating a cellular operator's rate of return. This was the topic of a heated debate during a recent proceeding at the Connecticut Department of Public Utility Control. DPUC Investigation Into the Connecticut Cellular Service Market and the Status of Competition, Docket No. 94-03-27.

[5] Cellular operators would nevertheless be required to comply with antitrust and fair business practices laws of general application.

[6] *E.g.,* Interim Opinion, Investigation on the Commission's Own Motion into Mobile Telephone Service and Wireless Communications, Decision No. 94-08-022, Investigation No. 93-12-007, 1994, Cal. PUC Lexis 487, *21 (Cal. PUC, Aug. 3, 1994) ("We elected this approach on the grounds that cellular service was 'discretionary' and that rapid technological change made industry oversight difficult and traditional service regulation problematic.").

[7] *E.g.,* Order Exempting Domestic Public Cellular Radio Telecommunications Service Providers from Regulation, In the Matter of Exemption of Domestic Public Cellular Radio Telecommunications Service Providers from Regulation Under Ch. 62 of the North Carolina General Statutes, Docket No. P.-100, Sub. 114, State of North Carolina Utilities Commission (Feb. 14, 1992) *aff'd* State of North Carolina v. North Carolina Cellular Association, Inc., No. 921OUC815 (N.C. Ct. App. 1993).

[8] Examples are Arizona, California, Connecticut, Massachusetts, and New York. The min-max ranges had been established when cellular was first introduced, usually based on estimates of what the market would bear, and were rarely changed. *E.g.,* Interim Opinion, Investigation on the Commission's Own Motion into Mobile Telephone Service and Wireless Communications, Decision No. 94-08-022, Investigation No. 93-12-007, 1994, Cal. PUC Lexis 487, *68 (Cal. PUC 1994). Thus, if they were ever based on the costs of the cellular carriers, they quickly became outdated. Some states (e.g., New York) allowed carriers to modify the ranges without cost justification. Others (e.g., Arizona) required rate cases to justify a modification of the minimum and maximum levels. However, because cellular carriers rarely maintain the expensive bureaucracies necessary to keep cost information in the manner generally required by regulatory bodies to justify rate changes, the min-max boundaries effectively became unchangeable in those states. Although the states would usually permit rate decreases, there was a fear on the part of the carriers that if costs rose in the future, they would never be permitted to increase any component of their rates. In effect, the regulation completely discouraged rate decreases.

[9] Wholesale rates are rates charged to resellers of cellular service. In contrast, retail rates are the rates actually charged to the end user. It might be expected that wholesale rates would set a floor below which retail rates would not fall, but this was not the case for all carriers. Moreover, regulation of wholesale rates does not put any cap at all on the price that can be charged to the end user. Thus, regulation of wholesale rates is not designed to protect the consumer of cellular service. Examples of states that regulated wholesale but not retail rates were Connecticut and Arizona.

[10] Report and Order, In the Matter of Bundling of Cellular Customer Premises Equipment and Cellular Service, at ¶ 11, CC Docket No. 91–34 (1992) (emphasis supplied).

[11] Omnibus Budget Reconciliation Act of 1993, Pub. L. No. 103-66, 107 Stat. 312 (1993). The Budget Act's state preemption provisions amended Section 332 of the Communications Act of 1934. 47 U.S.C. § 332(c)(3).

[12] Budget Act, Section 6002(c)(3)(A). It is not immediately clear what the purpose of subsection (ii) is. On its face it appears simply to impose the same burden of proof on a state as subsection (i), with the addition of showing that cellular is a substitute for landline service. One would expect a petitioning state to proceed under subsection (i), given that it imposes the lesser burden. The FCC has explained that, under its interpretation of the difference between the two subsections, where cellular is a substitute for landline.

> Congress merely intended to remove the presumption...that extant market forces are to be preferred over regulation as a means of ensuring just, reasonable, and nondiscriminatory CMRS rates. Put another way, where CMRS is the only available exchange telephone service, we construe [subsection (ii)] to mean that Congress' interest in promoting universal telephone service outweighs its interest in permitting the market for CMRS to develop in the first instance unfettered by regulation. [Footnote omitted.] As a practical matter, all this means is that concerns about anticompetitive conditions in the market for CMRS will be given greater weight where a state can show that such service is the sole means of obtaining telephone exchange service in a substantial portion of a state.

Report and Order and Order on Reconsideration, In the Matter of Petition of Arizona Corporation Commission, To Extend State Authority Over Rate and Entry Regulation of All Commercial Mobile Radio Services at ¶ 66, PR Docket No. 94-104 (1995). In that Report the FCC found that Arizona had not shown that cellular was a substitute for landline service under the statute. *Id.* at ¶ 67.

[13]   47 U.S.C. § 332(c)(3)(A); "State Wireless Regulation is Now at the Discretion of FCC," *State Tel. Reg'n Rep't*, Aug. 26, 1993.

[14]   Second Report and Order, In the Matter of Implementation of Sections 3(n) and 332 of the Communications Act: Regulatory Treatment of Mobile Services, at ¶ 252, GEN Docket No. 93–252 (1994).

[15]   Withdrawal of Petition, In the Matter of State Petition for Authority to Maintain Current Regulation of Rates and Market Entry Filed By the Public Service Commission of Wyoming, PR Docket No. 94-110 (1995).

[16]   Report and Order and Order on Reconsideration, In the Matter of Petition of Arizona Corporation Commission, To Extend State Authority Over Rate and Entry Regulation of All Commercial Mobile Radio Services, *supra* note 12, at ¶ 8. The first half of the FCC's separate decisions on the various state petitions are essentially the same; the FCC then proceeds to address the particulars of each state's petition in the second half of each one of the decisions.

[17]   *Id.* For example, a carrier might charge rates in excess of marginal cost, but it might be plowing that money back into the network to provide more and better service. *Id.* at ¶ 26.

[18]   *Id.* at ¶¶ 11–15, 22–25, 27. The FCC has denied petitions for reconsideration of its decisions in the California and Ohio decisions. Order on Reconsideration, In the Matter of Petition of the People of the State of California and the Public Utilities Commission of the State of California to Retain Regulatory Authority over Intrastate Cellular Service Rates, PR Docket No. 94-105 (1995); Order on Reconsideration, In the Matter of Petition of the State of Ohio for Authority to Continue to Regulate Commercial Mobile Radio Service, PR Docket No. 94-109 (1995). The state of Connecticut appealed the FCC's decision denying its petition for permission to continue regulation of cellular rates to the Court of Appeals for the Second Circuit. The FCC denied a motion to stay its decision pending appeal. Order, In the Matter of Petition of the Connecticut Department of Public Utility Control To Retain Regulatory Control of Wholesale Cellular Service Providers in the State of Connecticut, FCC 95-387, PR Docket No. 94–106, 1995. The Second Circuit upheld the FCC's denial of permission to continue rate regulation. Connecticut Department of Public Utility Control v. FCC, Docket No. 95-4108, slip opinion (2d Cir. March 22, 1996).

[19]   Licensing and Declaration Procedures, *supra* Ch. 6, note 12, at Annex One, pp. 4 (Belgium), 15 (Denmark), 22, 25 (France), 35 (Germany), and 68 (Italy).

[20]   Smith and Staple, *supra* Ch. 1, note 7, at xx.

[21]   Working Papers, *supra* Ch. 5, note 10, at 17; Iusacell Prospectus, *supra* Ch. 4, note 42, at 12. Some limited competition among operators on the price of telephone equipment and activation fees exists. *Id.*

[22]   *Id.* at 53.

[23]   Article 31 of the Telecommunications Regulations of Mexico.

[24]   Iusacell Prospectus, *supra* Ch. 4, note 42, at 53–54.

[25]   Section 4.5, *supra.*

[26]   At the time of this writing Mexico was still recovering from a plunge in the peso in December 1994, which affected the economy and the ability of its citizens to afford such products as cellular telephone service. "Grupo Iusacell Announces First Half Results," *PR Newswire Investorfax*, June 26, 1995.

[27]   Articles 60–63 of the Federal Law of Telecommunications, published in the Official Journal of the Federation on June 7, 1995; Leslie Crawford, "Rivals eager to enter Mexico's telecoms: The shape of market rules after deregulation in 1997 is becoming clearer," *Fin'l Times*, May 5, 1995.

[28]   Ch. 10, *infra.*

[29]   Request for Tenders, *supra* Ch. 6, note 17, at §§ 1.2.9, 3.1.2.

[30]   The study generally recommended that cellular service prices not be regulated, except in the situation "where tariffs are a major criterion in selecting a license holder. In that case, the tariff proposed, together with a mechanism to modify it in light of inflation, should be made a license condition." Licensing and Declaration Procedures, *supra* Ch. 6, note 12, at p. 87.

[31] Because CPP is the reverse of the common practice in the United States, the calling party receives a message advising him or her that he or she will be billed for the call and has the opportunity to hang up before charges are incurred. Moreover, a separate exchange is used for the mobile numbers of CPP subscribers.

[32] Alan Stewart and Alan Pearce, "PCS: first find your market," *Communications Int'l*, Oct. 1994, p. 78.

[33] Dianne Hammer, "Latin America's Cellular Growth Outstanding as Market Matures," *Radio Comm'ns Rep't*, Feb. 27, 1995 (describing move to CPP in Chile). The Jordanian tender specifies that CPP will apply, except when a customer is roaming outside his or her home area. § 1.2.7.

[34] Report and Order, In the Matter of Bundling of Cellular Customer Premises Equipment and Cellular Service at ¶¶ 9–10, CC Docket No. 91–34 (1992).

[35] Robert Patton, "Japan cellular activity heats up," Electronics, Jan. 9, 1995, p. 9; "Pacific Rim Expected to Outpace Cellular Growth Worldwide," *Mobile Phone News*, April 18, 1994; Leslie Helm, "Cellular Pact May Not Trigger Buy-U.S. Spree," *L.A. Times*, March 14, 1994, pt D, p. 1, col. 2.

[36] Telephone conversation with Peter Smith, World Bank, Feb. 13, 1995. Mr. Smith noted other reasons for the limited growth of cellular in Indonesia: a severe import tax, which has recently been repealed; poor quality of service; and government imposed restrictions on the number of customers that a cellular operator may have.

[37] Ray Heath, "Smart move by Smartcom," *S. China Morning Post*, Jan 26, 1995, p. 20.

[38] Telephone conversation with Peter Smith, *supra* note 36. As Mr. Smith noted, lower handset prices will not solve the teledensity problem unless activation rates are also maintained at an affordable level.

[39] Antitrust Law Developments (Third), Vol. I, (ABA 1992), pp. 131.

[40] A cellular operator is more likely to have market power in the service market if there are only two service providers. It is rare for a vendor of cellular phones to have market power in the equipment market because there are so many manufacturers and sellers of such phones.

[41] Antitrust Law Developments, Vol. I, pp. 132–36.

[42] Report and Order, In the Matter of Bundling of Cellular Customer Premises Equipment and Cellular Service, CC Docket No. 91-34 (1992).

[43] California was for years the last state in the United States to prohibit this practice. Cellular phones could not be discounted more than 20% or $10 below the wholesale price, whichever was higher. Peter Sinton, "How State Cellular Rule Has Failed," *S.F. Chronicle*, Dec. 7, 1994, p. D2. However, California finally decided in 1995 to allow bundling of cellular phones and service. "Industry Applauds California PUC Decision to Lift Ban on Cellular Tie-Ins," *Mobile Phone News*, April 17, 1995.

[44] Ian Channing, "Cellular Explosion; UK cellular telephone market," *Communications Int'l*, July, 1994, p. 10; "Germany Investigating Mobile Phone Marketing," *Newsbytes News Network*, May 9, 1994.

[45] "Competition is imperfect, says Hong Kong regulator," *FinTech Mobile Comm'ns*, May 19, 1994.

[46] Thus, it does not apply, for example, to such large landline carriers as GTE. The FCC originally applied the cellular separation rules to all wireline carriers, but subsequently limited the requirement to Bell companies. Notice of Proposed Rulemaking, In the Matter of Amendment of Section 22.901(c)(3) of the Commission's Rules, at ¶ 2 & n. 1, 1986 Lexis 3543 (1986).

[47] 47 C.F.R. § 22.903. Order, In the Matter of Bell South Corporation et al. Petition for Declaratory Ruling, DA 95-1401, ¶ 5 (1995). A large body of accounting practices, referred to as "Part 64 rules," has been developed for landline companies to use to calculate what they should charge their cellular affiliates. There is, however, a troubling question of what constitutes "joint marketing," which is prohibited, as opposed to "joint advertising," which is allowed. No satisfactory answer has been given on this point.

[48] Second Report and Order, In the Matter of An Inquiry Relative to the Future Use of the Frequency Band 806–960 MHz at ¶¶ 22–24, 46 F.C.C.2d, 30 R.R.2d 75 (1974); 47 C.F.R. §22.903(d).

[49] Report and Order, In the Matter of An Inquiry into the Use of the Bands 825–845 MHz and 870–890 MHz for Cellular Communications Systems, *supra* Ch. 3, note 53, at ¶¶ 48–49.

[50] Cincinnati Bell Telephone Co. v. FCC, Nos. 94-3701 et al. (6th Cir. Nov. 9, 1995).

[51] Licensing and Declaration Procedures, *supra* Ch. 6, note 12, at pp. 6, 51–52, 109.

[52] Overview, *supra* Ch. 3, note 56, at p. 5.

[53] "Grupo Iusacell News Release: Grupo Iusacell Files Antitrust Suit Against Teléfonos de México," Nov. 7, 1995; "Upstart Mexican Cellular Firm Alleges Telmex Tries to Stifle Competition," *Wall St. J.*, Nov. 8, 1995, p. A15.

CHAPTER 10

▼▼▼

UNIVERSAL ACCESS

In Chapter 2, the many benefits of telecommunications for developing countries were described. To realize fully these advantages [1] and to assure that they are fairly distributed, as many people as is reasonably possible should have access to telecommunications services. Universal access in the context of cellular has two components. The first is that the cellular network be built to ensure geographic coverage: the technical ability to obtain the signal needed to complete a call must be present. The second is that use of the system be affordable: the cost of the cellular phone and the cellular service must be within reach.

It is on the subject of universal access that the potential for divergence of the short-term goals of a private operator and the aims of the government is perhaps most apparent. The licensee, left to his or own devices, may choose to provide service only to the most profitable areas of a country. Accordingly, the government may want to condition the licensee's authorization to serve lucrative markets upon satisfaction of universal access obligations. As a philosophical matter, very few people would take the position that telecommunications service should not be subsidized for those who cannot afford. it. The tough questions are what subsidy mechanisms should be used and who should bear the cost. One participant in a World Bank symposium suggested that "[t]he question is not whether to subsidize but how to subsidize in order to derive the maximum benefit for the lowest cost" [2]. While the detailed specifics of any particular universal access plan are beyond the scope of this book, some discussion of general ideas is offered.

One way to accomplish universal access is to require the cellular provider (1) to engineer its network to ensure that nationwide geographic coverage is available and

(2) to make cellular service available at lower than market rates to those without the financial ability to pay full market rates. In the United States, as in most if not all countries, the landline carriers have these types of legal obligations to provide universal service. As part of the funding mechanism for universal service, the landline carriers receive money from support funds to contribute to their costs of providing universal service. The FCC has recently asked for comment on the extent to which wireless carriers should have universal service obligations if they are explicitly authorized to provide wireless in the local loop [3].

Another way to achieve universal access is to require the cellular operator (1) to provide the requisite geographic coverage and (2) to make contributions to a universal service fund that can be used by the government to assist those who need it. For example, the FCC recently proposed a plan under which certain subscribers would get a credit on their bills that they could direct to be paid to the local landline carrier or to some alternative provider of telecommunications service certified by the state and chosen by the subscriber as their carrier [4]. Or, rather than allowing each individual subscriber to choose their carrier, competitors could be required by the government to bid for the right to provide universal service to a defined group of subscribers [5]. Alternatively, the government might make prepaid phone cards, which a caller could use to complete local calls for up to a set time limit, available at a discounted price to those in need [6].

Regardless of how the universal service obligation is structured, it will be critical to ensure that the cellular provider will obtain a sufficiently high return on its overall investment that it will be willing to become a licensee. Each country must decide what the term "universal access" means to it and how the burdens of providing universal access are to be distributed among various types of telecommunications providers. The judgment required to ensure the rate of return necessary to attract providers to telecommunications services will be specific to each country and will depend on such factors as geography and demographics.

In developing countries, several definitions of "universal access" have been employed. The ITU has recommended that in rural areas a telephone be provided within one hour's walk for everyone [7]. Some countries have decided that each village with a population in excess of a certain number should have phone service [8]. Others have tried to identify multiple characteristics that will help them predict which areas of the country will benefit most from access to telecommunications services. For example, the benefit derived from the first pay phone installed might be different from the incremental value derived from an additional pay phone. Also, the presence of public institutions or productive enterprises in an area might indicate that a larger benefit would be derived from access to telecommunications services [9]. However, basing decisions on estimates of social and economic benefits to be derived raises a "chicken and egg problem": perhaps access to telecommunications services would draw public institutions and productive enterprises.

Some have suggested that service to rural areas, which typically involves a higher cost per subscriber for the telecommunications provider than service to urban or suburban areas, can be self-supporting [10]. Alternatively, the profits from cellular service to higher income markets may be sufficient to subsidize service to lower income markets. Another suggestion is that the cellular provider also be given the right to terminate international long-distance calls coming into the country and the right to receive half of the terminating charge [11]. This right could be quite lucrative under the current process of settlements for international calls. However, the international settlements process has been criticized because, among other things, the collection rates of foreign carriers are not cost-based and are high in comparison to U.S. rates, so that a high net outpayment deficit by U.S. carriers results. Thus, it may be that the international settlements process will be reformed so that the right to terminate international calls in a foreign country will not be so profitable in the future [12]. In anticipation of this, the right to receive a portion of the terminating international call revenue could decline over time [13].

Geographic radio coverage requirements are common conditions of cellular licenses and are often quite broad. Generally, major population, political, and business centers are expected to be covered first. For example, in Slovakia the cellular operator, Atlantic West B.V., was obligated to use its "best efforts" to construct at least one cell site in Bratislava within 12 months and at least one cell site in Kosice within 18 months [14]. Heavily traveled roads are usually another priority: not unexpectedly so, given the mobile nature of cellular communications [15]. After a period of a few years, carriers are often expected to cover significant percentages of the country in terms of both area and population. Percentage requirements as high as 94–100% are not unheard of, particularly in smaller countries [16].

It is interesting to contrast these requirements with those in the United States. The United States FCC formerly required that each cellular carrier must cover 75% of the area in its cellular geographic service area (CGSA). However, the CGSA is not the same as the metropolitan statistical area (MSA) or rural service area (RSA). MSAs and RSAs are the licensing areas designated by the FCC. The CGSA is that portion of the MSA or RSA that the carrier determines it wants to serve. If a part of the MSA or RSA is not designated by the carrier as being in its CGSA, or if a part of the designated CGSA is not covered within five years of the date that the carrier obtained its permit, the carrier may lose only the right to serve that area. But it does not lose its license and can continue to serve the remaining areas. No population coverage requirements have been imposed by the FCC, and the requirement to cover 75% of the CGSA has been eliminated [17]. However, a cellular operator must turn on its first cell site within 18 months of being granted a construction permit. Upon receiving notification that the first cell site in the MSA or RSA has been activated, the FCC will issue the operator its license [18].

The PCS rules impose stricter requirements. Licensees with 30-MHz blocks must provide service to at least one-third of the population within five years and two-thirds of the population within ten years. Failure to meet these requirements may result in forfeiture of the license [19].

The affordability problem can be dealt with by the government in different ways depending on how populous the locality is. (As an alternative to government regulation, residents of a particular area may enter into private agreements for phone sharing). In the most remote areas, government can provide cellular pay phones [20]. In somewhat more populous areas, the phones can be attended rather than coin-operated. In India, for example, the government established a program in which 200,000 phones were equipped with meters and put in the hands of entrepreneurs, frequently the handicapped, who set them up in bazaars, on street corners, or in cafes or shops whose owners feel they will attract more customers [21]. Also, community phone centers have been established in urban areas. These centers can be used by customers to make outgoing calls and receive messages, and are particularly valuable for job seekers [22].

One example of how cellular licensees have been required to satisfy both the geographic and affordability components of universal service is the case of South Africa. The 1993 invitation to tender issued by the South African government advised applicants for its two GSM licenses that fulfillment of community service obligations would be a key determinant that would weigh heavily in the balance in the selection process. The Pretoria government was under pressure from the African National Congress to take actions to correct the imbalance in telephone access between the black and white communities and to increase opportunities for entrepreneurship among blacks living in the townships [23]. (One in two white South Africans has access to a private telephone, but the figure is nearer to one in one hundred for the black population [24].) Both licensees, Vodacom and Mobile Telephone Networks, have community service obligations as part of their license terms [25].

The community service program being employed by Vodacom involves a franchising arrangement under two separate agreements: a leasing agreement by which Vodacom provides the franchisee with the necessary equipment in return for a monthly rental, and a franchising agreement under which the franchisee commits to provide the public with cellular phone and voice mail service in accordance with standards contained in the franchise contract. The equipment leased is a *phone shop*, which is a mobile *container* that can be relocated when necessary; up to ten cubicles equipped with phones (including transceivers) and metering units housed in the container; a central call charge management system for use of the center operator, to which all metering units and a printer are connected; the capacity to take advantage of an energy source (possibly solar); and an antenna. The franchisee purchases service from Vodacom and resells it to the public at a marked-up price, with a margin of about 33%. Vodacom selects the franchisees on the basis of potential entrepreneurial abilities (as demonstrated by creditworthiness, ability to pay a deposit toward the cost of the equipment, and ability to help identify a viable business site) and helps them obtain bank financing. (The equipment lease is assumed by a bank after six months, which gives the franchisee a chance to establish the business. The six-month period allows the franchisee to make the 10% down payment on the equipment in installments.) Vodacom also provides a manual to franchisees according to which they operate the phone shop. The first phone shop was opened in June 1994 [26].

Notes

[1] Each subscriber added to a telecommunications network increases the value of that network for existing subscribers. Henry Geller, "U.S. Telecommunications Policy: Increasing Competition and Deregulation," in Restructuring and Managing the Telecommunications Sector, supra Ch. 4, note 2, at 78.

[2] Timothy E. Nulty, "Emerging Issues in World Telecommunications," in *Restructuring and Managing the Telecommunications Sector, supra* Ch. 4, note 2, at 13. The idea of having a carrier charge below-cost rates to those of lesser means and charge those with the ability to pay (such as business people) rates that reflect the value to them of the service, so that the business as a whole makes money, is derived from an approach known as "Ramsey pricing." Those who have an ability to pay and for whom demand is price-inelastic (that is, their demand for a service does not decrease or decreases less than average when prices are raised) can be charged rates significantly above marginal cost. Saunders, *supra* Ch. 2, note 1, at 273–74.

[3] For a discussion of the universal service mechanism that applies to landline carriers, see, *e.g.,* Warren G. Lavey, "Universal Telecommunications Infrastructure for Information Services," 42 Fed'l Com. L.J. 151, 174 (1990); Notice of Inquiry, In the Matter of Amendment of Part 36 of the Commission's Rules and Establishment of a Joint Board, CC Docket No. 80-286 (1994) Lexis 4292 (1994). For the FCC's request for comments on universal service obligations for wireless carriers, see Notice of Proposed Rulemaking, In the Matter of Amendment of the Commission's Rules to Permit Flexible Service Offerings in the Commercial Mobile Radio Services, FCC 96-17, WT Docket No. 96-6 (released January 25, 1996).

[4] Notice of Proposed Rulemaking and Notice of Inquiry, In the Matter of Amendment to Part 36 of the Commission's Rules and Establishment of a Joint Board, ¶¶ 17–31 (1995).

[5] "FCC Chairman Hundt Outlines Five Ways to Ensure Successful Global Telecommunications Policy," *FCC News*, Nov. 7, 1995.

[6] "Development of Rural Telecommunications and the CTD," *supra* Ch. 2, note 4, at 19. Prepaid phones cards sold at full value have long been issued in Europe and Asia, and more recently in the United States, as a substitute for the cash needed to make long-distance calls. Card issuers can also make money by selling advertising space on the cards. *E.g.,* Harry Wessel, "Prepaid Phone Cards are Calling," *Orlando Sentinel*, Oct. 21, 1994; Donald Sabath, "Prepaid Cards can Offer Advertisers Phone Plug," *Plain Dealer*, Oct. 18, 1994, p. 2C. The government might then purchase cellular (or wireline) lines and equipment to serve as public phones.

[7] Luis G. Romero-Font, "The use of Cellular Radiotelephone Networks to provide Basic Exchange individual Line Telephone Service in Rural and Suburban areas and as an Emergency Back-up System to the Public Switched Network in cases of Natural Disasters," Telecommunications and Pacific Development: Alternatives for the Next Decade, Holland: Elsevier Science Publishers, 1988, p. 358.

[8] A. D. Clarkstone, B. E. Mrazek, and A. C. Diamond, "Rural Telecommunications Development in Botswana: Socio-Economic and Strategic Issues" in *Second International Conference on Rural Telecommunications,* Institution of Electrical Engineers, London, 1990 (describing Botswana project to bring phone services to villages of over 500 people).

[9] *Id.* See also Md. Towhidul Islam, "Modernization of Rural Telecommunications in Bangladesh - Problems and Solutions" in *Second International Conference on Rural Telecommunications,* Institution of Electrical Engineers, London, 1990(factors for prioritizing telecommunications projects, due to limited resources, were social benefit, economic benefit, technical suitability, political stability, and financial benefit to investor); Saunders, *supra* Ch. 2, note 1, at 128 (Costa Rican study indicated that presence of the following factors demonstrates that village has high-benefit potential: higher than average per capita income; relatively large population; location relatively far from major economic, social, and government center; above average educational level; populations clustered closely around telephone site) and 188 (discussing Vanuatu study that attempted to determine optimal public phone sites by quantifying losses resulting from lack of adequate communications by measuring time lost in travel to a telephone).

[10]   Ch. 3, note 51, *supra* and accompanying text (conversations with Ron Epstein).

[11]   Conversation with Ed Resor, Consultant, Oct. 3, 1994.

[12]   See Judy Arenstein, "Accounting Rates and The FCC's International Settlements Policy," in *1993 International Communications Practice Handbook*, 27–41.

[13]   Conversation with Ed Resor, Consultant, Jan. 3, 1995.

[14]   Operating License for 450 MHz Public Mobile Telephone Network, ¶ H/1 (Exhibit I2 to Joint Venture Agreement Between Post and Telecommunications Administration of the Slovak Republic). *See* Specification of Bases and Conditions For The International Public Bid For The Provision of Mobile Telephony Services in The Argentine Republic at § 5.5; United States Department of Commerce, Summary of Recent International Cellular Developments, Updated Feb. 26, 1994, at p. 3 (discussing requirements of Italian license); United States Department of Commerce, Summary of Recent International Cellular Developments, Updated Nov. 18, 1993, at p. 15 (discussing requirements of Philippine licenses).

[15]   *E.g.,* The Hashemite Kingdom of Jordan, Ministry of Post and Communications, the Telecommunications Corporation, Request for Tenders at ¶ 1.2.3; Licensing and Declaration Procedures, *supra* Ch. 6, note 12, at Annex One, p. 23 (French GSM license), p. 51 (Greek GSM licenses), and p. 89 (Portuguese GSM licenses).

[16]   Licensing and Declaration Procedures, *supra* Ch. 6, note 12, at p. 14 (GSM operators in Denmark obliged to cover 95% of area and 98% of population), p. 35 (one German GSM operator offered to cover 94% of population and that condition was written into its license), and p. 89 (GSM operator in Portugal obliged to cover 99% of the population); United States Department of Commerce, Summary of Recent International Cellular Developments, Updated, Feb. 26, 1994, at p. 5 (Singapore requires "island-wide coverage").

[17]   Report and Order, In the Matter of An Inquiry Into the Use of the Bands 825–845 MHz for Cellular Communications Systems; and Amendment of Parts 2 and 22 of the Commission's Rules Relative to Cellular Communications Systems, 86 F.C.C.2d 469, at ¶ 97, 49 R.R.2d 809 (1981); 47 C.F.R. §§22.946, 22.947.

[18]   47 C.F.R. § 22.946.

[19]   47 C.F.R. § 24.203.

[20]   *E.g.,* Asia-Pacific Cellular Operators: 1994, § 2.12 (Northern Business Information)(describing rural telecommunications project in Thailand that plans to establish, among other things, twenty thousand TDMA cellular pay phones).

[21]   Sam Pitroda, "Development, Democracy, and the Village Telephone," *Harv. Bus. Rev.*, Nov. 1993/Dec. 1993, p. 86.

[22]   Saunders, *supra* Ch. 2, note 1, at 24 (describing Community Telephone Centers set up in peripheral neighborhoods of Lima, Peru). Bell Atlantic has voluntarily set up a similar arrangement for homeless persons in some metropolitan areas it serves. E.g., "Bell Atlantic Prepares the Disadvantaged for New Lives with Hopeline Program," *Land Mobile Radio News*, Oct. 29, 1993 (indicating $3,000 per market cost to establish the service, which is offered free of charge).

[23]   Jim Ayre, "Cellular Issue Resolved in South Africa," *Africa Communications*, Feb. 1994, p. 16; United States Department of Commerce, Summary of Recent International Cellular Developments, Updated Nov. 18, 1993, p. 17.

[24]   Nick Cottam, "Phones ring change in black townships," *Sunday Times*, June 5, 1994.

[25]   Vodacom is owned by Telkom SA Ltd, the national telephone company (50%), Vodaphone (35%), and Rembrandt, a South African tobacco company (15%). MTN is owned by M-Net (25%), Cable & Wireless (25%), Naftel (20%), Fabcos (5%), Transtel (20%), and a pension fund affiliated with the Congress of South African Trade Unions (5%). "Phones ring change in black townships," supra note 24. According to the World Bank, Vodacom has committed to provide 22,000 community service lines while MTN's commitment is 7,500. Draft memorandum of Mohammad Mustafa, Sept. 28, 1994.

[26]   Draft memorandum of Mohammad Mustafa, Sept. 28, 1994; "Phones ring change in black townships," *supra* note 24.

CHAPTER 11

▼▼▼

SERVICE QUALITY

11.1 INTRODUCTION

The quality of the cellular radio service provided by an operator can be judged objectively according to quantitative standards. These standards are likely to be of two types: network quality and customer service. This chapter will discuss some of the standards that are typically used [1]. Such standards may be imposed by the government or voluntarily adopted by cellular licensees. Other countries, such as Mexico, have simply mandated that the licensee establish a service quality standards system without specifying the standards [2]. Provisions for compensation to users when the carrier fails to meet these standards may also be established. Whether a country should impose mandatory standards or leave quality judgments up to the provider probably should, like price regulation, turn in large part on whether cellular is a substitute for wireline service and whether competition is present.

11.2 NETWORK QUALITY

Network quality, or the quality of a call on the cellular system, "is linked to the probability that the customer can initiate, receive, and maintain an acceptable call in the service area" [3]. Based on this definition, three extremely important indicia of service quality are ineffective attempts, lost calls, and distortion (the quality of the transmission). Measurements of call quality are made during the *busy hour*, usually coinciding with the time of heaviest commuting, because this represents the worst-case scenario [4].

The percentage of ineffective attempts is generally an indication of whether the capacity of the network is sufficient. When a call in progress is lost, it may be—but is not necessarily—an indication of whether a cell's signal strength is high enough. Whether signal strength is high enough may depend on the kind of equipment the customer is using. Handheld portable units operate at a lower power than mobile and transportable units, so that a call can be more difficult to commence and maintain (hand off from cell to cell) with a portable phone [5].

It is important to set standards for both ineffective attempts and lost calls because one statistic may be inflated at the expense of the other. For example, one might engineer the network so that it will allow a caller to commence a call even when signal strength is quite weak. This will reduce ineffective attempts, but will probably also increase lost calls.

Some countries of the European Union impose legal limits on how many ineffective attempts and lost calls a cellular system can have. However, the United Kingdom decided not to impose grade of service quality standards, believing that no reliable methodology exists [6].

The United States FCC does not impose mandatory standards for grade of service, because it wanted the interplay of market forces to determine the grade of service. It feared that mandatory standards could "have the detrimental effect of denying service to economically marginal markets" [7]. Nevertheless, cellular carriers in the United States use the quality standards discussed in this section to judge their performance on an internal basis. The FCC does set certain technical standards to which a cellular system must adhere and does require that cellular systems be interconnected with the landline network [8].

The third measurement, distortion, can be caused by such things as noise that occurs during a call, including static and crosstalk. Distortion is also referred to as *signal to noise and distortion* (SINAD).

In addition to these three most important measures of network quality, others can be used. Access timing—the number of seconds a cellular caller must wait between when he or she presses the "send" button on the phone and when the call is connected at the other end—can be measured. The quality of a cellular network's service may also be ascertained by outages of either cells or switches. For example, Greece requires service without interruption, but allows six hours per week of outage to perform maintenance [9]. An outage may be the result of any one of several causes. It may be due to failure of the cellular equipment or failure of electrical power or landline telephone service to the switch or cell site. That is, inability to complete calls may or may not be the result of action or inaction by the cellular operator. (However, the adverse effects of power failures can be avoided through the use of emergency backup generators.) Obviously, measurement of outages overlaps with the measurement for ineffective attempts. If the system is out, the attempt to make a call will not be effective.

Most of these measurements can be recorded in two ways: internally and externally. Internally, the cellular system maintains a switch-based record of certain call statistics [10]. Thus, a government could require the cellular operator to provide reports on every call made on its system during a particular period. Externally, regulators could outfit a vehicle with cellular phones and test equipment and drive around the service area, making calls and taking measurements [11]. System-generated (internal) numbers are more valid statistically because they measure all calls made during a particular time frame. However, they are subject to manipulation by the operator. External measurements can be statistically valid if sufficient calls are made. A combination of internal and external measurements would ensure that all criteria are being met and would allow each type of measurement, internal and external, to act as a "sanity check" on the other.

A combination of internal and external measurements would also be desirable because of differences in the way compliance with certain criteria is measured. Take, for example, the situation of a call that cannot go through or that is cut off because the landline carrier's service to the cellular network is interrupted. A switch would not consider this an ineffective attempt or lost call, but the external tester (and the customer) would. In this case, the external tester's result is more reliable. Conversely, consider the situation of a tester who loses a call because the cell-site power is too low. Whether the call is lost when the tester is riding on a little-traversed back road or a busy highway, each lost call would count the same to an external tester when he or she is calculating the percentage of the calls he or she made that were lost. Yet the call lost on the busy highway would be of more concern from a quality control point of view, because a weak cell site on a busy highway is worse than one on a little-traveled road. The switch data takes this into account because it counts every call made by every vehicle, and so will count a higher percentage of calls being lost. In this case, the switch's count is more reliable.

11.3 CUSTOMER SERVICE

Cellular licensees could also be expected to meet certain customer service standards. Examples of such standards commonly include the following [12]:

- A customer's phone should be activated within a certain period of time.
- A toll-free and airtime-charge-free number should be available for customer service, possibly on a 24-hour per day, 7-day per week basis.
- Incoming calls to customer service should be answered within a certain period of time.
- Customer service centers must be reasonably available.
- Certain features (such as call forwarding, free emergency calls, and call waiting) should be provided.

Adherence to these standards could be verified based on records maintained by the licensee, by customer service surveys, or by government testing. The government may want to contract the testing function to outside firms [13]. The regulatory agency could also impose requirements for satisfactory resolution of customer complaints filed with it.

One further criterion that a government might use to evaluate a cellular operator's performance is penetration. As noted in connection with the discussion of whether the cellular operator should be private or public (Chapter 4, *supra*), penetration is used as a measure of success in the cellular industry. Penetration requirements have also been imposed on landline service providers in developing countries [14]. The government could set what it believes to be reasonable levels of penetration, or ask prospective licensees to propose such levels and use their commitments as a basis for selection.

Notes

[1] Most network equipment manufacturers provide quality standards as well. Balston & Macario, *supra* Ch. 1, note 1, at 98 (discussing quality standards for Nordic Mobile Telephone). Recommended standards have been developed for GSM equipment. Such standards are a measure of the level at which the system can be expected to perform. The recommended GSM standards have been adopted as license conditions by some countries.

[2] The 1989 amendment to the 1957 concession granted to Iusacell's predecessor requires such a system to be established and updated every two years.

[3] Balston & Macario, *supra* Ch. 1, note 1, at 252.

[4] Busy hour for Bell Atlantic is usually around 4–5 p.m. on a weekday. Conversation with Hans Leutenegger, Jan. 3, 1995.

[5] Balston & Macario, *supra* Ch. 1, note 1, at 252–53.

[6] Licensing and Declaration Procedures, *supra* Ch. 6, note 12, at Annex One, pp. 15, 24, 35, 51, 89, and 108.

[7] Report and Order, In the Matter of An Inquiry Into the Use of the Bands 825–845 MHz and 870–890 MHz for Cellular Communications Systems, *supra* Ch. 3, note 53, at ¶ 95.

[8] *Id.* at ¶¶ 84–94. See 47 C.F.R. §§ 22.933 (specifications for cellular equipment compatibility), 22.917 (emission limits), and 22.915 (modulation requirements).

[9] Licensing and Declaration Procedures, *supra* Ch. 6, note 12, at p. 54.

[10] An analog switch cannot measure distortion. A digital switch can measure bit error rates, which provide some indication of distortion. Discussion with Hans Leutenegger, Jan. 11, 1995.

[11] For example, Bell Atlantic conducts external testing every month in every region by driving a van with four phones and making continuous calls. Two phones make calls on the Bell Atlantic system and two on the competitor's system. Of the two phones on each system, one makes three-minute calls and one makes sixty-minute calls. The phones are operated by computer and they call computers located at the regional headquarters. The phones on either end send 1-kHz tones to each other during the call. The computers measure ineffective attempts, lost calls, and access timing. They also compare distortion of the tone with the undistorted tone sent as a measure of call quality. *Id.*

[12] *E.g.,* Andrew Burroughs, "MobiLink; B-side brand," *Cellular Marketing,* Aug. 1993, p. 40; The Hashemite Kingdom of Jordan, Ministry of Post and Telecommunications, The Telecommunications Corporation, Request for Tenders at § 4.4.

[13] Smith and Staple, *supra* Ch. 1, note 7, at xvi.

[14] Telephone conversation with Andrew Fyfe, Price, Waterhouse, Washington, D.C., Jan. 18, 1995.

CHAPTER 12

▼▼▼

LICENSE REVOCATION

12.1 INTRODUCTION

When a licensee fails to comply with legal standards governing its conduct, it may be sanctioned. While development of a complete regulatory regime is beyond the scope of this book, certain conduct that is sufficiently serious to warrant revocation of a cellular license will be addressed. Other less serious penalties, not discussed here, could include short-term renewal of a license (a sort of probationary period), a conditional license (conditional on correcting some unsatisfactory circumstance), a fine, or an admonishment.

12.2 UNITED STATES

The United States Communication Act, originally enacted in 1934, does not have a section dealing specifically with the revocation of cellular licenses. When cellular was introduced, the Federal Communications Commission (FCC) decided to apply a combination of legal provisions: some that had initially been passed to govern the provision of telephone service, and others that had been aimed at radio and television broadcasting. Cellular systems, like satellite systems, are considered by the FCC to use radio stations because they engage in radio communications. Two sections of title 47 of the United States Code (the title that deals with telecommunications) address revocation of licenses: Section 312, which is concerned with revocation generally, and Section 159, which covers revocation for failure to pay regulatory fees.

As can be seen from some of its language, Section 312 was originally aimed at broadcasters [1]:

The Commission may revoke any station license or construction permit:

1. For false statements knowingly made either in the application or in any statement of fact which may be required pursuant to section 308;
2. Because of conditions coming to the attention of the Commission which would warrant it in refusing to grant a license or permit on an original application;
3. For willful or repeated failure to operate substantially as set forth in the license;
4. For willful or repeated violation of, or willful or repeated failure to observe any provision of this Act or any rule or regulation of the Commission authorized by this Act or by a treaty ratified by the United States;
5. For violation of or failure to observe any final cease and desist order issued by the Commission under this section;
6. For violation of Section 1304, 1343, or 1464 or title 18 of the United States Code; or
7. For willful or repeated failure to allow reasonable access to or to permit purchase of reasonable amounts of time for the use of a broadcasting station by a legally qualified candidate for Federal elective office on behalf of his candidacy.

The word "willful" is defined as "the conscious and deliberate commission or omission of [an] act, irrespective of any intent to violate any provision of this Act or any rule or regulation of the Commission authorized by this Act or by a treaty ratified by the United States" [2]. The word "repeated" is defined as "the commission or omission of such act more than once or, if such commission or omission is continuous, for more than one day" [3].

Before revoking a license or permit under Section 312, the FCC must follow certain procedures. The Commission must serve an order on the licensee to show cause why the license should not be revoked. The order must contain "a statement of the matters with respect to which the Commission is inquiring" and must give the licensee 30-days notice to appear before the Commission and give evidence. If the Commission decides, after hearing or waiver of hearing, to revoke the license, its revocation order "shall include a statement of the findings of the Commission and the grounds and reasons therefore..." [4]. The Commission bears "both the burden of proceeding with the introduction of evidence and the burden of proof..." [5].

U.S. cellular licenses have rarely been revoked. Revocations occur where the rule violations are particularly egregious and are accompanied by concealment of the wrongdoing. In a recent case, In re Applications of Algreg Cellular Engineering, et al. [6], a number of licenses were revoked under Section 312(a)(2), quoted above, because the principals of the licensees were found to have engaged in a "risk-sharing" scheme in violation of the FCC's lottery rules and to have concealed their violations from the Commission [7]. Some of them were also found to have violated other rules of the Commission by making false certifications in their applications [8].

The statutory provision that a license may be revoked for a reason that would be grounds for denying it in the first place provides further insight, because denials of cellular licenses are more frequent than revocations. One important instance of grounds for denial arises out of the Communication Act's requirement in Section 310(d) that control over a license may not be transferred without prior consent from the FCC. This covers both legal changes (such as changes in who owns a controlling block of the licensee's stock if the licensee is a corporation) and *de facto* changes (such as an abdication of responsibility for the system). The FCC will deny a license where the license applicant relinquishes *de jure* or *de facto* control over the system to another party without such consent [9]. However, the FCC seems reluctant to revoke a license for rule violations that would have served as grounds to deny the license unless the violations were intentionally deceptive and accompanied by concealment [10].

Further guidance can be obtained from the broadcasting area. The FCC has said that its policy statement regarding character qualifications in broadcast licensing provides guidance in common carrier cases [11]. More broadcast than cellular licenses have been revoked under Section 312. As with cellular, broadcasting licenses are generally not revoked for rule violations unless they are intentional and the licensee makes misrepresentations to the FCC [12]. However, there are exceptions [13].

A license may also be revoked for failure to pay regulatory fees under Section 159 of title 47 (Section 9 of the Communications Act) [14]. Regulatory fees are assessed by the FCC to recover costs incurred in carrying out its enforcement activities, policy and rulemaking activities, user information services, and international activities. In a revocation proceeding under this section of the Communications Act, the licensee bears the burdens of introduction of evidence and of proof. A hearing is required only where the licensee's response to an order for failure to pay regulatory fees raises a substantial and material question of fact. A revocation order for failure to pay regulatory fees will not become final until the licensee has exhausted its right to judicial review [15].

12.3 EUROPEAN UNION

Some grounds for revocation of cellular licenses are fairly typical among the countries of the European Union. Several are similar to grounds stated in United States law: failure to fulfill the terms of the license (Belgium, Denmark, Germany, Greece, and the Netherlands), failure to pay fees (Belgium, Denmark, Greece, and the United

Kingdom), and change in ownership (similar to a legal change in control in the United States) (Denmark and the United Kingdom) [16].

Unlike the United States, however, several EU countries (Denmark, Greece, the Netherlands, and the United Kingdom) will revoke a cellular license when the licensee enters bankruptcy, liquidation, or receivership [17]. This difference may be a result of a distinction in the respective uses of term *bankruptcy* in the U.S. and European legal systems. In the United States, bankruptcy encompasses both reorganization and liquidation, whereas bankruptcy often does not connote reorganization in European countries. A company that is reorganized under Chapter 11 of the U.S. bankruptcy code would be unlikely to emerge from bankruptcy court protection as a viable business if its FCC licenses were revoked automatically [18].

Various other grounds for revocation exist under the laws of some of the EU countries. These include failure to eliminate interference caused by the licensee's station (Belgium); change in the composition of capital of the operator (France); the general economic and social interest, safety of the State or public order, and the disappearance of "the grounds on which the license was based" (the Netherlands); and failure to comply with an official order or if a change of shareholding would affect national security or relations with a foreign government (the United Kingdom) [19].

12.4   JORDAN

The Jordanian invitation to tender that the IFC (World Bank) assisted in preparing similarly provides for revocation in case of bankruptcy, liquidation, or receivership; failure to pay various fees due under the license; failure to provide agreed coverage; or violation of the terms of the license or related agreements. Revocation may occur upon 30-days notice, although cure periods are allowed for failure to pay fees and general violation of terms [20].

Notes

[1]   47 U.S.C.S. § 312(a) (1994). Section 308 of title 47 requires written applications for licenses. Sections 1304, 1343, and 1464 of title 18 make criminal the broadcasting of lottery information; fraud by wire, radio, or television; and broadcasting obscene language, respectively.
[2]   47 U.S.C.S. § 312(f)(1).
[3]   47 U.S.C.S. § 312(f)(2).
[4]   47 U.S.C.S. § 312(c).
[5]   47 U.S.C.S. § 312(d).
[6]   Decision, CC Docket No. 91-142 (1994). The most recent decision in the case is that of the FCC's Review Board, which upheld the 1993 decision of the Administrative Law Judge to revoke the licenses and construction permits involved and denied the applications for authorizations that had not yet been granted. The decision of the Review Board is being appealed to the Commissioners. If the licensees lose at that level, they may appeal to the Court of Appeals for the District of Columbia Circuit.

[7] The FCC's rules prohibited (1) partial settlements among lottery applicants; (2) any applicant in a cellular lottery from having an interest in more than one lottery application in any given market; and (3) alienation of interests in lottery applications. The licensees had engaged in a scheme under which the lottery winner would convey an interest in its profits to each of the other parties to the agreement. All of the parties to the agreement were lottery entrants. The risk-sharing agreement violated all three of the prohibitions. *Id.* at ¶ 32.

[8] *Id.,* ¶ 33. The Review Board found that false certifications by themselves would not have been sufficient basis for revocation. *Id.* at ¶¶ 77–78.

[9] See Telephone and Data Systems, Inc. v. FCC, 19 F.3d 42, 48–50 (D.C. Cir.1994) *on remand,* 9 FCC Rcd 7108 (1994), *Memorandum Opinion and Order on Reconsideration,* FCC 96-3 (Released Jan. 23, 1996). The FCC has incorporated the Act's requirement into its rules at 47 C.F.R. § 22.137.

[10] Memorandum Opinion and Order and Notice of Apparent Liability for Forfeiture, In re Catherine L. Waddell, 8 F.C.C. Rcd 2170, 72 R.R.2d 500 (1993); Order, In re Randolph Cellular Limited Partnership, 7 F.C.C. Rcd 2114, 1992 Lexis 1486 (1992); Memorandum Opinion and Order, In re Application of Century Cellnet of Jackson MSA Limited Partnership, 6 F.C.C. Rcd 6150, 70 R.R. 2d (1991).

[11] Memorandum Opinion and Order, and Notice of Apparent Liability, In re Applications of David A. Bayer, 7 F.C.C. Rcd 5054, 71 R.R.2d 308, note 18 (1992).

[12] Opinion, In the Matter of David R. Price, 7 F.C.C. Rcd 1838, 70 R.R.2d 803 (Rev. Bd. (1992); Brian C. Murchison, "Misrepresentation and the FCC," *Fed. Com . L.J.,* Vol. 37, 1985, p. 403.

[13] Opinion, In the Matter of Silver Star Communications - Albany, Inc., 65 R.R.2d 761 (Rev. Bd. 1988) (licenses revoked for violation of FCC's distress sale rule, even where no actual misrepresentation or lack of candor issues were tried in the case, because the violations exposed the licensee as "utterly untrustworthy to hold the subject licenses"). The revocation was reduced to a fine on appeal. Memorandum Opinion and Order, In the Matter of Silver Star Communications—Albany, Inc., 70 R.R. 2d 18 (1991).

[14] 47 U.S.C. § 159(c)(3).

[15] Report and Order, In the Matter of Implementation of Section 9 of the Communications Act; Assessment and Collection of Regulatory Fees for the 1994 Fiscal Year at ¶¶ 61 et. seq., 9 FCC Rcd 5333 (1994).

[16] Licensing and Declaration Procedures, *supra* Ch. 6, note 12, at Annex One, pp. 10, 16, 37, 52, 82, and 109.

[17] *Id.* at pp. 16, 52, 82, and 109.

[18] However, the U.S. FCC would not allow a bankrupt licensee to transfer the license without its consent, as would normally occur in liquidation, where the debtor's assets are sold. For this reason a cellular licensee in the United States may not give a security interest in its cellular license, although it may give a security interest in the proceeds of the sale of such a license. *E.g.,* In re Ridgely Communications, Inc. 139 B.R. 374 (Bankr. D. Md. 1992). A security interest in a license could result in an involuntary transfer of control without FCC approval.

[19] Licensing and Declaration Procedures, *supra* Ch. 6, note 12, at pp. 10, 24, 82, and 109.

[20] Request for Tenders, *supra* Ch. 6, note 17, § 3.20.

CHAPTER 13
▼▼▼

INTERCONNECTION

13.1 INTRODUCTION

Unless a cellular system is interconnected with other networks, including the landline telephone system, cellular customers will only be able to call and receive calls from other cellular subscribers. A delay in establishing interconnection for a new entrant into the cellular market can be a tremendous advantage for a preexisting cellular carrier [1]. Moreover, interconnection costs are usually a substantial percentage of a cellular carrier's expenses. In EU countries, for example, interconnection is usually responsible for about one-third of a cellular company's expenses [2]. Excessive interconnection costs can mean the difference between a profitable and a money-losing venture for the cellular carrier. The importance of interconnection is thus apparent.

It has sometimes been difficult for cellular carriers to obtain landline interconnection. Accordingly, regulatory authorities have required landline telephone companies to allow cellular carriers to interconnect. Moreover, such interconnection must be available on a nondiscriminatory basis. Nondiscrimination is particularly important when the landline company is also a cellular service provider because of the possibility that it may favor its cellular affiliate over that affiliate's competitors. Another issue that has arisen is whether the rates and other terms and conditions on which interconnection is available, even when nondiscriminatory, are reasonable.

There is also a question of whether the landline company should pay the cellular company for interconnection. This could be of particular interest to developing countries, where the cellular company may at some point have as extensive a network as the landline company. An outgrowth of this question is whether cellular operators should have an obligation to interconnect with other telecommunications providers.

13.2 UNITED STATES

The United States FCC generally requires landline telephone companies to provide the type of interconnection requested by the cellular carrier and to provide nonwireline (A side) cellular carriers with interconnection that is equal in type, quality, and price to that provided to their wireline (B side) counterparts [3]. To facilitate enforcement of the equal connection requirements, the Commission has required cellular companies owned by the regional Bell holding companies to operate as fully separate subsidiaries. In the context of discriminatory interconnection, the FCC has stated that [4]:

> a separate cellular entity greatly simplifies the opportunity of other cellular operators to gain interconnection rights to the landline network on the same basis as the telephone subsidiary offering the underlying basic cellular transmission facilities. Requiring a wireline carrier to offer cellular service only through a separate entity with facilities separate from the wireline service reduces the opportunity that 'technical complexity' or similar reasons could be invoked by wireline carriers as a basis for denying interconnection.

Disputes have also arisen about whether charges for and other terms and conditions of interconnection, even if nondiscriminatory, have been reasonable [5]. The FCC requires that rates be reasonable. Two nonprice requirements are also imposed. Interconnection must be of the type requested by the cellular carrier. Moreover, nonwireline carriers are entitled to interconnection that is different from that provided to the wireline cellular carrier, as long as the request is reasonable [6].

However, the FCC has so far left the specific amount of compensation and most other terms and conditions to negotiations between the landline companies and the cellular carriers. In 1994, it issued a notice of proposed rulemaking on the issue of whether landline interconnection rates for cellular carriers and new PCS licensees (collectively, along with paging companies and certain other wireless service providers, commercial mobile radio services (CMRS)) should be tariffed. In requesting parties to present views on the respective advantages and disadvantages of tariffed as opposed to negotiated interconnection, the FCC set out its preliminary thoughts on the subject [7]:

> A comparison of the two approaches reveals that both have benefits and costs. Tariffing, with its attendant filing and reporting requirements, could impose administrative costs upon carriers, which could lead to in-

creased rates, but there are also transaction costs in developing negotiated arrangements. Moreover, a tariffing requirement might lead to less
flexible interconnection arrangements for CMRS providers. Tariffs may
not provide sufficient flexibility for crafting multiple options that reflect
the different needs of different carriers.

On the other hand, tariffing is an established mechanism for ensuring
that rates, terms and conditions are reasonable and that carriers do not
engage in unreasonable discrimination. Because LECs [local exchange
carriers] have some incentive to delay or impose barriers to the development of competition from new CMRS services, such as PCS, a tariff
process might be in the public interest.

The FCC requested comment on two potential substitutes for tariffs: the mandatory inclusion of a most favored nation clause in all interconnection agreements and a
requirement that all interconnection agreements be filed with the Commission [8].

Cellular operators have also argued that interconnection compensation should
flow both ways. The idea behind compensation is that the cellular carrier pays the
landline carrier because the landline carrier completes calls from the cellular to the
landline network. Historically, the cellular carriers have been treated as end users
(customers) rather than cocarriers in this respect. That is, cellular carriers pay the
landline companies for completing calls from the cellular network, but the landline
companies do not pay the cellular carrier for completing calls from the landline to
the cellular network. However, cellular carriers have argued that they should be
treated as cocarriers and allowed to charge the landline carriers for landline calls
terminated to the cellular network. The issue is referred to as *mutual compensation*,
and the FCC has recently said that the principle shall apply to interstate transmissions. Under this principle, the landline carriers would compensate the cellular carriers for "the reasonable costs incurred by such providers in terminating traffic that
originates on LEC facilities" [9]. The FCC proposed that on an interim basis the
principle of mutual compensation be implemented by way of a *bill and keep* policy.
Under this arrangement, both the LEC and the cellular carrier would charge a rate of
zero for the termination of traffic. Cost-based pricing options are also being considered for the short and long terms [10].

Given the growth of wireless providers and their insistence on being treated
as carriers entitled to mutual compensation, it was perhaps inevitable that someone
would suggest that they too have an obligation to allow interconnection with their
own wireless networks. However, the FCC has decided that it will not order broad
mandatory interconnection at this time. The FCC expressed its general preference to
leave such matters up to the market, to be realized primarily through private negotiations and arrangements. It also stated that it felt it would be premature to impose
an interconnection requirement on CMRS providers given the rapidly changing state
of wireless technology. Concern about the need for wireless customers to be able to
reach each other was also alleviated by the fact that all CMRS end users can cur-

rently connect with users of any other network through the LEC landline network. Nevertheless, it expressed concern that some CMRS providers (particularly those affiliated with landline LECs) might obtain market power and use interconnection rates and terms and conditions to raise rivals' costs in an anticompetitive manner. Accordingly, the FCC relied on the fact that CMRS providers are common carriers and as such must provide communications service on reasonable request to hold that it may order specific interconnection arrangements with respect to particular carriers if it believes it to be in the public interest after opportunity for hearing. It also reminded CMRS providers that they may not discriminate unjustly or unreasonably with respect to interconnection [11].

13.3 EUROPEAN UNION

As in the United States, issues of nondiscriminatory and reasonable interconnection have arisen in the countries of the EU. On occasion, protracted disputes have arisen, sometimes delaying the introduction of competition in cellular [12]. This section will discuss first the law and problems of interconnection in these countries prior to the issuance of the EU's 1994 Green Paper on mobile services. It will then review the positions of the Mobile Green Paper and the subsequent CEC communication on the subject of interconnection.

Several EU countries, such as Denmark, France, Germany, Greece, the Netherlands, Portugal, and the United Kingdom have explicitly required the landline telephone company to provide interconnection to cellular operators, frequently on a nondiscriminatory and reasonable basis. Often, interconnection rates and other terms are to be negotiated between the parties, sometimes with the assistance of the telecom ministry if agreement cannot be reached. This has been the case in France, Germany, the Netherlands, and the United Kingdom [13]. The process has thus been not dissimilar to that which has prevailed in the United States.

In some countries, agreement on interconnection has been resolved fairly quickly. More frequently, however, interconnection has not been any easy issue to settle. The second GSM operator in Germany was engaged in a long and well-publicized dispute over interconnection to the landline telephone network. Difficulties have also been encountered in Greece. To some degree, the problem can be avoided by publishing the rates and other terms of interconnection in advance of bidding on the license [14].

The EU's Mobile Green Paper recommended that interconnection be available to cellular operators in a nondiscriminatory and cost-based fashion. Specifically, it stated that "interconnection conditions...must be set on the basis of objective criteria, be transparent and non-discriminatory, cost-oriented, and compatible with the principle of proportionality, as well as respecting essential requirements" [15]. The need for transparency was said to encompass transparency concerning accounting practices where an operator has both a fixed and a mobile network. Transparency was said to imply that information about interconnection be available to the applicable national regulatory agency and the EU [16]. The significance of mandating cost-based rates that can be verified by audit practices is that this requirement can prevent anticompetitive behavior by the landline telephone company. Specifically, it will reveal whether the landline carrier is inflating interconnection rates to all cellular carriers on the theory that it does not mind overcharging its own cellular business because that is simply a transfer of money from one pocket to another [17].

The Mobile Green Paper also concluded that terms of interconnection should be left to commercial agreement. Thus, disputes will continue to be left to the national regulatory authorities to be resolved, with the concomitant possibility that the delay inherent in the process can be used as a stalling tactic by a landline carrier that is also in the cellular business or that sees cellular as a competitor to its landline business. Perhaps the number of disputes will diminish as the negotiation process becomes more regularized.

The communication from the Commission of European Communities to the Council and Parliament, which followed from the Green Paper, listed as an area for priority action the development of a regulatory regime for interconnection based on the EU's open network framework. This is a general telecommunications reform that would include interconnection with the landline system by mobile operators as well as others. This regulatory regime, it said, should give priority to commercial agreement between the parties under the supervision of regulatory authorities and strict scrutiny under EU competition rules as set out in the Treaty of Rome. The regulatory authorities should lay down the principles for negotiation, and common principles should be applied for interconnection charges [18]. (For a discussion of the ability to use alternative methods of interconnection in the EU, see Chapter 8.)

The Mobile Green Paper's recommendations did not address mutual compensation: whether landline carriers should compensate cellular carriers for calls originated on the landline network and completed on the cellular network. Nor did it discuss whether cellular providers should be required to provide each other direct interconnection with their networks. However, the report underlying the Mobile Green Paper did recommend such a requirement, and the United Kingdom already imposes one [19].

Notes

[1] "Philippine Telecommunication: Deregulation Crowds the Field," *East Asian Executive Rep'ts*, April 15, 1994; "German cellular interconnection row settled next year," *FinTech Mobile Comm'ns*, Nov. 8, 1990.

[2] Licensing and Declaration Procedures, *supra* Ch. 6, note 12, at pp. 28, 47.

[3] The Need to Promote Competition and Efficient Use of Spectrum for Radio Common Carrier Services, 59 R.R. 2d 1275 (1986); 2 FCC Rcd 2910 (1987).

[4] Report and Order, In the Matter of An Inquiry Into the Use of the Bands 825–845 MHz and 870–890 MHz for Cellular Communications Systems, *supra* Ch. 3, note 53, at ¶ 50.

[5]   Interconnection charges can be calculated in a variety of ways and are highly dependent on the type of technology in use. *E.g.,* "BT's price cap may be undermined by new interconnection rules," *FinTech Telecom Markets,* July 22, 1993. Several sources are available on the subject of interconnection. *E.g.,* George Calhoun, *Wireless Access and the Local Telephone Network,* Norwood, MA: Artech House, 1993.

[6]   Notice of Proposed Rule Making and Notice of Inquiry, In the Matter of Equal Access and Interconnection Obligations Pertaining to Commercial Mobile Radio Services at ¶ 103, 9 F.C.C. Rcd 5408 (1994). Initially, the FCC did not specify exactly what type of interconnection should be provided, but after disputes on the issue arose, the FCC required the provision of so-called "Type 2" interconnection. Type 2 interconnection is more favorable for the cellular carriers than Type 1 interconnection. Some landline telephone companies had provided only Type 1 interconnection. For a discussion of the difference between Type 1 and Type 2 interconnection, see Kennedy, *supra* Ch. 4, note 12, at 40–41.

[7]   Notice of Proposed Rulemaking, *supra* note 6, at ¶¶ 115–116.

[8]   *Id.* at ¶¶ 119–120.

[9]   Second Report and Order, *supra* Ch. 9, note 15, at ¶ 232.

[10] Notice of Proposed Rulemaking, In the Matter of Interconnection Between Local Exchange Carriers and Commercial Mobile Radio Service Providers, FCC 95-505 (Released January 11, 1996).

[11] Second Notice of Proposed Rulemaking, In the Matter of Interconnection and Resale Obligations Pertaining to Commercial Mobile Radio Services at ¶¶ 28-44, CC Docket No. 94-54, 1995 FCC Lexis 2662 (1995).

[12] Nor are these sorts of problems limited to the cellular sphere. *E.g.,* Richard L. Hudson, "Telesystem's Fight With Telecom Italia Reflects Hurdles to Opening EU Market," *Wall St. J.,* Nov. 29, 1994, p. A9C (describing problems encountered by an alternative provider of landline service when attempting to obtain leased lines from the state-owned telephone company).

[13] Licensing and Declaration Procedures, *supra* Ch. 6, note 12, at Annex One, pp. 15, 21, 25, 36, 52, 82, 90, and 111.

[14] *Id.* "German cellular interconnection row settled next year," *FinTech Mobile Comm'ns,* Nov. 8, 1990, at pp. 21, 36–37, 52, 90.

[15] Overview, *supra* Ch. 3, note 56, at p. 11. The principle of proportionality "requires that substantial public regulation should be limited to only those mobile systems providing services to the general public." *Id.* at p. 5. That is, private mobile radio systems are not included, except to the extent needed to avoid harmful radio interference and ensure frequency efficiency.

[16] *Id.* at p. 11.

[17] Licensing and Declaration Procedures, *supra* Ch. 6, note 12, at p. 51.

[18] CEC Communication, *supra* Ch. 7, note 9, at 8, 35, 53.

[19] Licensing and Declaration Procedures, *supra* Ch. 6, note 12, at pp. 8–81 and Annex One, p. 11. The obligation of a cellular carrier to interconnect in the U.K. extends as well to any public-switched network, including cable systems.

CHAPTER 14

▼▼▼

RESALE

14.1 INTRODUCTION

Resellers are companies that pay wholesale rates to purchase telecommunications services from the licensees that operate the networks. Resellers then mark up the price of the service and sell it to their subscribers at prices that they hope are competitive with those that the licensees charge their own end users. Resellers bill their customers and maintain their own customer service organizations. They may package cellular service with other telecommunications products and services.

Regulatory authorities often require cellular operators to sell service to resellers, and to do so on a nondiscriminatory basis. Governments promulgate resale requirements with respect to network operators in the hope that resellers will provide additional competition to (and thus impose market discipline on) the licensees. This chapter will examine the types of requirements that have been imposed on cellular licensees with regard to resale.

14.2 UNITED STATES

The Federal Communications Commission's policy on resale predates the cellular industry. The FCC has held that landline common carriers may not unjustly or unreasonably restrict the resale of certain services under Section 201(b) of the Communications Act and that they may not unjustly or unreasonably discriminate among

---

resellers (e.g., a carrier may not favor its own resale operations over a reseller that is unaffiliated with the carrier) under Section 202 of the Act. Resale was defined as "an activity wherein one entity subscribes to the communications services and facilities of another entity and then reoffers communications service and facilities to the public (with or without adding value) for profit." Several public benefits were identified by the FCC as resulting from the elimination of restrictions on resale. First is the provision of communications services at rates more closely reflecting costs, because carriers must avoid being undercut in their pricing by resellers. Second is better management of specialized communications networks, expected to result from the need of resellers to offer extensive advice and assistance to customers to make their services attractive and responsive so that they can compete with facilities-based carriers. Third is the avoidance of waste of communications capacity. The final anticipated benefit is the creation of additional incentives for research and development of ancillary devices to be used with transmission lines [1].

The FCC adopted this policy for the cellular industry early on. It found that the same justifications for eliminating resale restrictions that applied to landline service also applied to cellular, although it said that it was "not certain that true resale of cellular service [would] develop" [2]. An example of a restriction that the FCC has found unreasonable in the cellular industry is a carrier-imposed requirement that resellers offer that one carrier's service exclusively [3]. Thus, the reseller cannot be prevented from buying and reselling both the A and B side service in a market.

The law prohibits restrictions and discrimination with respect to basic services, but arguably not as to enhanced cellular services. The FCC has made a distinction between basic and enhanced services in the landline context [4]:

> [E]nhanced service shall refer to services, offered over common carrier transmission facilities used in interstate communications, which employ computer processing applications that act on the format, content, code, protocol or similar aspects of the subscriber's transmitted information; provide the subscriber additional, different, or restructured information; or involve subscriber interaction with stored information. Enhanced services are not regulated under title II of the Act.

Of course, the fact that the FCC does not regulate enhanced services under title II of the Communications Act does not mean it might not decide that it may regulate them under another provision of the law. Moreover, the FCC has not specifically designated particular cellular services as enhanced, and thus not subject to regulation.

Not only has the FCC required cellular carriers to sell capacity to pure resellers, it has insisted that a carrier must sell to its facilities-based competitor—the other cellular carrier in the market. This requirement originally developed primarily because most wireline cellular carriers brought their networks online before the nonwireline carriers, and the FCC sought to minimize the effects of this "head start" on the development of competition. However, the FCC has modified that policy so that a carrier need not sell to the competing cellular carrier after the competitor has had its license for five years. After that period of time, the competitor is expected to be fully operational. Requiring resale to the competitor after that time might even be contrary to the public interest. For example, the competitor might try to avoid the expense of converting to digital by reselling the other carrier's service, thus delaying the benefits of digital service. Also, it might open the possibility of collusion between the two carriers as to coverage areas [5].

Unlike some other countries (such as the United Kingdom), the FCC has not required cellular carriers to establish separate resale subsidiaries. Thus, operators of cellular networks do not need to create an affiliate to engage solely in resale activities and do not need to establish a wholesale rate to charge themselves [6]. Regardless of whether they choose to establish such a separate subsidiary, however, licensees may not "foreclose similarly-situated customers from ordering the same or substantially similar service under the same terms and conditions or...[fail] to provide such service under the same terms and conditions" [7].

Pricing issues have long been a sore spot with resellers in the United States, who have claimed that facilities-based carriers unfairly subsidize what they believe are money-losing retail operations with wholesale profits. They feel that it is such alleged cross-subsidization that makes it impossible for them to compete with the carriers. However, the fact that the FCC does permit carriers to apply volume pricing discounts if they can be justified as reasonable, and resellers have not had operations large enough to take advantage of these discounts, may provide an explanation for why resellers have not been more financially successful. The recent entry by MCI into the cellular resale business will provide an opportunity to test this theory. MCI has entered into resale agreements with several major cellular operators that will let it offer cellular service to 75% of the U.S. population, including the top 100 markets. It has also purchased Nationwide Cellular Service, Inc., a reseller with 275,000 customers, for $190 million, or about $690 per subscriber [8].

The FCC has tentatively concluded that PCS licensees who provide commercial mobile radio services (CMRS) should also be obligated to permit resale. However, it has indicated that it will probably impose a time limitation on the obligation to resell to a facilities-based competitor. This limitation would be comparable to the five-year limit in the cellular context described above, but may be a shorter or longer time frame. It has also sought comment on whether number portability should be required. Number portability would allow a customer of a PCS licensee that resells a cellular carrier's service to migrate to the PCS system when it comes online and still keep the phone number he or she was assigned when a cellular customer. However, the FCC tentatively rejected a proposal made by resellers that licensees be required to allow resellers to insert their own switches into the licensees' network [9].

14.3 EUROPEAN UNION

Historically, few member nations of the EU imposed resale obligations on cellular licensees. The United Kingdom is one of the exceptions. Other countries, such as France, also have "service providers," a term often used in Europe to refer to resellers, although the terms of the relationship are left to commercial negotiation. Germany left the matter of service providers up to the marketplace for the GSM licensees but obligated the PCS (DCS 1800) licensee to establish resale relationships with service providers [10].

During the initial stage of introduction of wireless service, licensees in the United Kingdom were not permitted to sell airtime directly to users. Until January 1, 1994, they could sell only to unaffiliated service providers. After that date they could establish separate affiliated companies to buy and resell service. These related service providers were, however, allowed to purchase only under the same terms and conditions as nonaffiliated resellers. Licensees are bound to connect with any service provider operating approved equipment as long as it is competent and demands service from the licensee solely for resale [11].

Cross-subsidization of the service provider with revenues from the mobile licensee is generally prohibited in the United Kingdom, although exceptions have been made. New entrants have been allowed to cross-subsidize for one year, and extensions to two years have been granted. Upon expiration of these exemptions, separate accounting rules will go into effect for the previously exempt companies [12].

The number of service providers in the United Kingdom has declined over the past several years. In 1990, there were 60; as of 1995, there were 38. This reflects both a number of business failures and a series of mergers and takeovers [13].

The report commissioned by the EU in preparation for its 1994 Green Paper on mobile services recommended the institution of several legal requirements with respect to resale. It suggested a prohibition on exclusive dealing clauses by which a licensee prevents service providers from buying and reselling the cellular service of other licensees. It felt that such a prohibition would be consistent with similar controls on dominant wholesalers in other industries under national and EU law. The report also advocated that resellers have the right to trade in any member state of the EU. It proposed rejection of qualifications for service providers that are not based on basic principles of financial and technical competence and imposition of a requirement that operators show good cause as to why a particular service provider should not be accepted. Finally, it recommended that regulatory agencies enforce open, transparent, and arms-length relationships between network operators and their service providers [14].

As summarized in the Overview, the conclusions of the EU's Mobile Green Paper with respect to service provision were as follows [15]:

1. Commercial freedom should be guaranteed, allowing the provision of services by independent Service Providers, as well as direct service

provision by mobile network operators. All existing restrictions in licenses impeding such activity should be abolished.

2. Commercial freedom should include the opportunity for service providers, whether independent or integrated into or forming part of mobile network operations, to offer a combination of services provided under different mobile licenses, as well as the ability to provide services in different Member States, subject only to the provision of the Treaty [of Rome] competition rules.

3. Service Providers should not be subject to licensing procedures and may be subject only to a requirement for declaration of their activities to the National Regulatory Authority(ies) of the Member State(s) where they choose to operate.

4. A Code of Conduct for Service Providers should be established. The Code should, in particular, identify on the basis of voluntary participation by Service Providers, measures to safeguard essential requirements and commitments with regard to permanence of services, availability, and quality of service. It should also provide guidelines concerning technical, financial, and commercial practices in the sector, consistent with competition rules.

5. Mobile network operators should, in line with their obligation to provide open, transparent and non-discriminatory conditions for interconnection...have an obligation to accept all reasonable requests by Service Providers to deal, within the limits of normal commercial practice and Community competition law (including requests from Service Providers integrated into other mobile network operation). It should be possible to challenge any refusal to deal before the National Regulatory Authority.

6. In order to guarantee open, transparent, and non-discriminatory conditions for independent Service Providers, mobile network operators should be required by their licence to provide for sufficient transparency, in particular concerning their accounting practices, to allow supervision of the service provision activities integrated into their operations.

7. The commercial relationship established between Service Providers and mobile communications operators should be subject to full mutual recognition by Member States. No restrictions should be applied concerning any activity resulting from such relationships in one Member State on the activity in any other Member State.

The provision of services in the context of cross-border roaming agreements represents the provision of a service by independent Service Providers or by Service Providers integrated into a mobile network operation in the territory of a Member State other than the Member State in which the original commercial activity was established. Such activity should not be subject either to any restriction or to any surcharge or equivalent measure unrelated to the cost of the provision of the roaming facility itself, whether imposed as a result of regulatory or other action.

The communication to the European Parliament and Council on the Consultation on the Mobile Green Paper proposed measures with respect to service providers. There was a public consensus that service providers are an important way of providing telecommunications services. However, there was disagreement on whether operators should be required to deal with service providers, or whether they should be free to select their own means of distribution. There was also a lack of consensus on whether service providers that are affiliated with operators should be required to be fully separate. However, the CEC found strong support for a code of conduct for service providers. Such a code would set out the rights and duties of service providers with regard to both the operator and the end user. Formulation of such a code was denominated a priority action for the mobile sector. A decision was not made as to whether the code should be mandatory or voluntary [16].

Notes

[1]   Report and Order, In the Matter of Regulatory Policies Concerning Resale and Shared Use of Common Carrier Services and Facilities, 60 F.C.C.2d 261 (1976).
[2]   Report and Order, In the Matter of An Inquiry Into the Use of the Bands 825–845 MHz and 870–890 MHz for Cellular Communications Systems, *supra* Ch. 3, note 53, at ¶¶ 103–107. The policy is incorporated into the FCC's regulations governing cellular. 47 C.F.R. § 22.903.
[3]   TRAC Communications v. Detroit Cellular Telephone Co., 5 F.C.C. Rcd. 4647 (1989).
[4]   47 C.F.R. § 64.702(a).
[5]   Report and Order, In the Matter of Petitions for Rule Making Concerning Proposed Changes to the Commission's Cellular Resale Policies, CC Docket No. 91-33 (1992).
[6]   Cellnet Communication, Inc. v. FCC, 1992 U.S. App. LEXIS 12964 (1992).
[7]   *Memorandum Opinion and Order,* In the Matter of Cellnet Communications Inc. v. Detroit SMSA Ltd Partnership, File No. E-91-95, at ¶ 26 (1994).
[8]   "MCI signs with cellular," *FinTech Telecom Markets,* Aug. 17, 1995; Tom Dellecave, Jr. and Mary E. Thyfault, "Wireless Maneuvers—AT&T, MCI, and Sprint prepare for increased competition," *Informationweek,* Aug. 14, 1995, p. 24; Linda Kay Sakelaris, "Leaders Prepare to Step Ahead in Evolving Mobile Marketplace," *Radio Comm. Rep't,* July 24, 1995, p. 25.
[9]   Second Notice of Proposed Rulemaking, *supra* Ch. 13, note 11, at ¶¶ 83–98.
[10]   Licensing and Declaration Procedures, *supra* Ch. 6, note 12, at pp. 15–16, 31–32, and Annex One, pp. 25, 35, 41, 112.
[11]   *Id.* at Annex One, p. 112.
[12]   "UK telecoms regulator extends Hutchison's Orange cross-subsidy of DSB to 2 years," *AFX News,* Aug. 15, 1995; "C & W's Mercury Wins Right to Cross-Subsidise in Mobile Telecoms for Further Year," *AFX News,* Feb. 16, 1995.
[13]   Michael Dempsey, "Trade body challenges regulator," *Fin'l Times,* Nov. 27, 1995, Special Section on Mobile Communications, p. 2, col. 1.
[14]   Licensing and Declaration Procedures, *supra* Ch. 6, note 12, at pp. 52, 88, and 108.
[15]   Overview, *supra* Ch. 3, note 56, at pp. 9–10.
[16]   CEC Communication, *supra* Ch. 7, note 9, at 6–7, 18–19, 28, 30–31, 53.

PART III

▼▼▼

SATELLITES

CHAPTER 15

▼▼▼

TECHNOLOGY

15.1 INTRODUCTION

The technology of satellite systems can be divided into two parts: the space segment and the Earth segment. The discussion of the former will cover certain characteristics of the satellites that are placed into space, such as where in space they are located, how many satellites are needed to produce the desired coverage area, and what frequencies the system uses. The Earth segment refers to what sorts of services can be provided to users on the ground and what kinds of stations are employed to take advantage of them. Discussion of the Earth segment will also cover the linkages between the parts of a satellite system. It is worth reiterating that the focus of this book is on two-way, point-to-point services and the discussion omits other types of satellite-based services (such as broadcasting).

15.2 SPACE SEGMENT

15.2.1 The Geostationary Orbit

In 1945, the science writer Arthur C. Clarke suggested that the entire globe could be served by three satellites strategically placed in the *geostationary orbit* (GSO). The GSO is a circular orbit above the equator at an altitude of 22,300 miles or 35,800 kilometers. A satellite in the GSO will orbit the Earth in a time of 23 hours, 56 minutes, or approximately one day. Hence, satellites in the GSO are synchronous with the Earth's rotation and appear to observers on Earth to be at a fixed point in the sky (i.e., geostationary). The fact that the geostationary satellite is continuously visible from any point within its potential service area (about 40% of the Earth's surface) has been especially significant because it means that fixed ground antennas do not need to be continually reoriented to track the satellite. Therefore, the cost of computer-controlled tracking equipment has been avoided [1].

One problem with geostationary satellites (also referred to as GEOs) is that because their orbit is so high, there is a delay in propagation of their signals [2]. Anyone who has carried on an intercontinental telephone conversation using satellite transmissions can understand how annoying the delay is, as it often leads to parties talking over each other because each starts his or her sentence at the same time as the other.

The geosynchronous approach presents other problems when it comes to serving the higher latitudes (those further from the equator). This is because the "look angle" from the satellite to the Earth's surface becomes lower and lower, making effective communications increasingly difficult. The look angle is the angle at which a ground transceiver "looks" to a satellite to which it is operational [3]. The higher the look angle, the less the likelihood that the signal will be blocked by buildings or terrain.

Commercial geostationary satellites currently provide fixed satellite service in the C band and Ku band of the radio spectrum. The Ku band is also used by some geostationary satellites to provide certain mobile services [4]. The distinction between mobile and fixed satellite services is discussed in Section 15.3. Numerous proposals are pending at the ITU and FCC for fixed and mobile service from geostationary satellites in the Ka band [5].

The allocations for the C, Ku, and Ka bands are somewhat different for various regions of the globe. For the United States, the allocations are shown in Table 15.1.

TABLE 15.1
C-, Ku-, and Ka-Band Allocations for the United States

Band	Uplink Frequency	Downlink Frequency
C	5.925–7.075 GHz	3.7–4.2 GHz
Ku	14–14.5 GHz	11.7–12.2 GHz
Ka	27.5–31 GHz	17.7–21.2 GHz

Source: [6].

Each band has certain advantages and limitations. Use of the C band preceded use of the Ku and Ka bands because the C-band allocation is shared with terrestrial microwave and components were thus already available at these frequencies. But because of this sharing, the power that can be generated by a satellite using the C band is restricted to avoid interference with microwave transmission. Therefore, receiving antennas must be large and located outside of metropolitan areas. However, C-band transmissions are almost unaffected by atmospheric conditions, whereas the Ku and Ka bands can be seriously attenuated by rain. There are hybrid satellites that use both the C and Ku bands. It is expected that the use of the C band for new satellites will decline as the industry moves to the Ku and, further in the future, the Ka bands [7].

Some GEOs utilize the L and S bands (1–3 GHz) of the spectrum. These are used to provide mobile satellite services.

15.2.2 Low-Earth, Middle-Earth, and Highly Elliptical Orbits

Like geostationary satellites, the remaining types of satellites can also be classified by their orbits. The highly elliptical orbit (HEO) is a specialized orbit in which a satellite continuously swings very close to the Earth, loops out into space, and then repeats its swing-by. The middle-Earth orbit (MEO) is a circular orbit approximately 5,000 to 10,000 miles above the surface of the Earth, not necessarily above the equator. The low-Earth orbit is, like the MEO, a circular orbit that need not be equatorial, but that stretches approximately from 100 to 1,000 miles above the Earth's surface [8].

The MEO is in some senses a compromise between the lower orbits and the geosynchronous orbit. A MEO system involves more delay and higher power levels than satellites in the lower orbits. However, a MEO system requires fewer satellites to achieve the same coverage [9].

Highly elliptical orbits are designed to give better coverage to countries with higher northern or southern latitudes. Systems can be designed so that the apogees are arranged to provide continuous coverage for a particular area [10]. The apogee is the point in the orbit where the satellite is farthest from the Earth.

The U.S. FCC calls these three orbits highly elliptical orbits, medium-altitude orbits, and lower-altitude orbits and collectively refers to satellites in them as low-Earth-orbiting satellite systems (LEOs) [11]. This book will generally adopt the FCC's terminology and will henceforth use the acronym LEO to cover all three, although when specific services or systems are discussed, some distinctions may be made among the three.

At the FCC, LEO satellites have been further subdivided into Big LEOs, which operate at frequencies above 1 GHz, and Little LEOs, which operate below 1 GHz. Little LEOs provide data services, travel in the lowest orbit, and require relatively small amounts of spectrum. Big LEOs provide voice and data services, may travel in any of the three nongeosynchronous orbits, and require much more spectrum [12].

Many entities are now planning to exploit LEOs for commercial use. Until the late 1980s, a satellite in any orbit other than the GSO had to be tracked continuously from the ground, which required expensive tracking equipment located at Earth stations. The tracking stations picked up each satellite as it rose above the horizon and passed overhead and then handed off the satellite to the next tracking station as it fell below the horizon. However, technological advances have made it possible for a satellite dipping below the horizon to hand off its signals to another satellite [13].

LEOs have been allocated parts of four bands of the radio spectrum. They utilize or expect to utilize portions of the VHF (30 MHz–300 MHz), UHF (300 MHz–1 GHz), and L and S bands (1–3 GHz). At least one proposal has been made to the U.S. FCC for a LEO system in the Ka band [14].

Satellites traveling in a lower orbit than the GSO have certain advantages and disadvantages resulting from their proximity to Earth. One advantage is that transmission delays can be eliminated. The satellites can be made lighter and cheaper than geostationary satellites and can provide for the ability to have small, lightweight, and inexpensive receive antennas on the ground. However, many more satellites are needed to provide continuous communications with lower orbit satellites than with geostationary satellites. Moreover, LEOs require more fuel in order to maintain altitude, due to the stronger effects the Earth's gravity has upon them. The projected life of a LEO ranges from about 5 to 15 years, whereas GEOs may last from 10 to 15 years. The combination of large satellite constellations and frequent replacement requires inexpensive, timely, and reliable launch capabilities [15].

15.2.3 Launching

The subject of launching of satellites deserves a brief mention for several reasons. First, it is a not insignificant part of the expense of putting a satellite system in working order. Second, launch capacity is not always easy to obtain. Third, there have been a number of unsuccessful launches.

The FCC has estimated that the cost of launching a satellite into the geostationary orbit is 20 times that of launching a LEO [16]. One source put the cost of GEO launches (comprising the cost of the launch vehicle and launch insurance) at about $60–$65 million for a C-band satellite and $74–$111 million for a Ku-band satellite [17].

Until the 1980s, commercial launches in the United States were conducted exclusively by the federal government, which would contract with the launch provider in order to provide the launch service to the satellite owner. However, a European-backed French company, Arianespace, began to emerge as a contender for a large part of the global commercial space launch market. Congress then passed the Commercial Space Launch Act [18] in order to encourage a domestic commercial launch industry. Until the explosion of the space shuttle *Challenger*, NASA also was permitted to use its shuttle fleet for commercial launches; after the explosion, such use was prohibited in the absence of a specific need [19]. Due to the lack of U.S. development of rockets to launch satellites and the cost and limited availability of the space shuttle, satellite operators have recently begun using rockets developed in other countries. However, the U.S. government has limited the number of American satellites that can be launched on foreign rockets [20].

There are only a limited number of rockets that are capable of putting GEOs into orbit. These include the Ariane, the Atlas (U.S.–Lockheed Martin), the Proton (Russia), and the Long March (China) [21]. The Delta rocket (U.S.–McDonnell Douglas) is also able to launch some GEOs [22].

There are a greater number of launch vehicles capable of putting LEOs into orbit. These launch vehicles can be smaller than those for GEOs, given that the satellites themselves are smaller. New rocket designs will attempt to put a half-dozen or more satellites into different orbits with a single launch [23].

Unfortunately, there have been a number of unsuccessful launches (or as one participant euphemistically referred to them, "launch anomalies") of both GEOs and LEOs [24]. The success of the various rockets are tracked by insurance companies. One recent survey shows rates ranging from 83.3–100%, depending on the rocket [25]. Another source, in late 1993, estimated that historically, approximately 15% of all commercial GEO launches have resulted in total or constructive total loss [26]. In response to this phenomenon, Arianespace in 1995 decided to offer free relaunch on a priority basis in event of failure. Companies would still be responsible for the cost of replacement and launch insurance [27].

15.3   EARTH SEGMENT

15.3.1   Mobile Satellite Service

The term *mobile satellite service* (MSS) has been defined by the United States FCC as "radiocommunication service (1) between mobile Earth stations and one or more space stations; or (2) between mobile Earth stations by means of one or more space stations" [28]. Mobile Earth stations are also referred to as *individual user units*.

There are three types of MSS: aeronautical mobile satellite services (AMSS), maritime mobile satellite service (MMSS), and land mobile satellite service (LMSS) [29]. Mobile satellite services can be provided by both GEOs and LEOs.

Land mobile satellite service combines both a huge growth market for investors and tremendous potential for developing countries. Different types of satellites can provide different mobile services.

Little LEOs will provide nonvoice mobile services such as facsimile transmission, e-mail, and position-finding. (Some of these services, such as e-mail, may also be provided on a fixed basis.) Positioning data may be used to help rescue teams locate emergency sites quickly. They may also be used in business to monitor the movements of trucks, boats, planes, shipments, or other assets [30]. The United States FCC also refers to Little LEOs as *NVNG MSS* (for nonvoice, nongeosynchronous mobile satellite systems).

Some Little LEO systems provide services on what is called a *noncontinuous* or *store and forward* basis. This is accomplished where the system has insufficient satellites to cover simultaneously the entire service area. For example, one or two satellites might be launched, so that each part of the service area can communicate with the satellite system for a specified period (as little as 20 minutes) every day. During that time, e-mail messages, for example, can be transmitted between the satellite and the ground station.

Big LEOs and GEOs will provide mobile voice services and are also capable of providing the same services as Little LEOs. It is in the provision of these voice services that they will both compete directly and be integrated with cellular systems. It has been suggested that cellular systems will draw customers who are interested in local or regional coverage, while satellite systems will attract those who are willing to pay more for international or global service on a single handset. However, some MSS systems are being designed to accommodate dual-mode handsets: phones that will use cellular service if it is available and satellite service if it is not.

Satellites that provide mobile voice service can, like cellular systems, also provide an alternative to the landline network in currently unserved areas. Some companies that are in the process of creating MSS systems have indicated that they hope they will be able to serve developing countries. They have made suggestions that resemble some of the ways in which cellular has been used. Examples that have been put forth include solar-powered phone booths that could be subsidized by the government and franchised kiosks for towns of 5,000–20,000 inhabitants that may or may not be subsidized. The question will be whether the prices charged will be within the range of affordability in such countries [31].

Mobile satellite services are typically accessed by use of a transceiver (transmitter/receiver), which is the individual MSS user unit or mobile Earth station. The most obvious example of a transceiver is the mobile phone unit. Some mobile user units only transmit or receive. For example, when geopositioning service is provided, the transmitter and receiver are separate. The transmitter is carried with the person (such as a hiker in a remote area) or thing (such as a stolen car) being tracked and emits a radio signal that is picked up by satellites. The satellites then transmit to the geopositioning receiver, which calculates the transmitter's location [32].

A part of any satellite system ground segment (either mobile or fixed) is the tracking, telemetry, and control (TTC) equipment. This is used to follow the path of the satellites and redirect them if necessary.

In addition to the individual MSS user unit and TTC equipment, the ground segment for mobile satellite service includes a gateway Earth station. The gateway station interconnects the mobile satellite system with other communications networks or with other user units. The gateway station communicates with the individual MSS units via satellite.

Whereas the radio connection between the MSS user unit and the satellite is a *service link*, the connection between the satellite and the gateway station is a *feeder link* (so called because it "feeds" information from other networks). As the FCC has said, "[a]n MSS service link allocation without appropriate feeder link allocations would be analogous to an automobile without an engine" [33]. However, feeder links do not need to be in the same spectrum band as the service links. Indeed, feeder links are considered by the FCC to be in the fixed satellite service [34].

15.3.2 Fixed Satellite Service

The term "fixed satellite service" was defined by the ITU in 1971 as [35]:

A radiocommunications service:

- between Earth stations at specified fixed points when one or more satellites are used; in some cases this service includes satellite to satellite links, which may also be effected in the intersatellite service;
- for connection between one or more Earth stations at specified fixed points and satellites used for a service other than the fixed satellite service (for example, the mobile satellite service, broadcasting satellite service, etc.).

Thus, FSS is provided between satellites and Earth stations that are "at specified fixed points" and may or may not be used in conjunction with MSS.

The FCC has defined Earth station complexes and their functions broadly enough to encompass those used in both the fixed and mobile services [36]:

The term communication satellite Earth station complex includes transmitters, receivers, and communications antennas at the Earth station site together with the interconnecting terrestrial facilities (cables, lines, or microwave facilities) and modulating and demodulating equipment necessary for processing of traffic received from the terrestrial distribution system(s) prior to transmission via satellite and of traffic received from the satellite prior to transfer of channels of communication to terrestrial distribution system(s).

The communication satellite Earth station complex interconnects with terminal equipment of common carriers or authorized entities at the interface; accepts traffic from such entities at the interface, processes for transmission via satellite, and performs the transmission function; receives traffic from a satellite or satellites, processes it in a form necessary to deliver channels of communication to terrestrial common carriers or such other authorized entities, and delivers the processed traffic to such entities at the interface.

Today there are over 200 geosynchronous satellites providing FSS. Originally, these satellites were typically used for long-distance or overseas telephone service: a way to connect different wireline systems separated by great distances or bodies of water [37]. Since then, other fixed satellite services have been developed. Broadcast, while not the main focus of this book, is one of the more economically significant FSS applications. Two other uses are very small aperture terminal (VSAT) network services and specialized private network services. VSAT networks are typically used by large corporations and organizations to connect a central location such as a headquarters with hundreds or thousands of field sites. Small Earth stations with antennas 1–2 meters in diameter are installed at the field locations, and a somewhat larger hub station is required at the central site. Satellites are used to connect the central and field locations, allowing for data, voice, and video distribution and/or two-way transmission. Specialized networks refer to private communications networks that use nonstandard transmission formats, usually higher data rates. For example, such a system is used for purposes of printing the *Wall Street Journal* at several locations throughout the United States [38].

Fixed satellite services can also provide basic telecommunications services to rural and remote areas where there are low concentrations of traffic, making wireline systems uneconomical [39]. This has been one of the fundamental goals of INTELSAT, a global treaty organization that operates a number of satellites. (INTELSAT is discussed in detail in Chapter 18.) As with cellular, the declining cost of the technology (both for the space and ground segments) has made the use of satellites to deliver basic telecommunications services to the populations of developing countries more prevalent, especially for long-distance service, because the cost of satellite service within a given satellite's footprint is insensitive to distance. However, in most instances satellites have not proven as economical as cellular [40].

Use of the fixed satellite service as a substitute for landline service can be expected to expand in the future. For example, the owners of Teledesic, which has proposed an 840-satellite LEO system that would cover 95% of the Earth's surface, have offered to dedicate some of the capacity of their network on a nonprofit basis for developing countries' needs, such as education and health care [41].

Notes

[1]   Rita Lauria White and Harold M. White, Jr., *The Law and Regulation of International Space Communication*, Norwood, MA: Artech House, 1988, p. 9 [hereinafter "R. White"].

[2]   Adamson, *supra* Ch. 2, note 32, at 128.

[3]   For fixed satellite services, the look angle is generally not lower than 5 degrees and for mobile satellite services it is usually no lower than 30 degrees. Joseph N. Pelton, *Wireless and Satellite Communications: The Technology, the Market, and the Regulations,* Englewood Cliffs, NJ: Prentice-Hall, 1995, pp. 51, 257 [hereinafter "Pelton"].

[4]   Adamson, *supra* Ch. 2, note 32, at 228, 234.

[5]   *E.g.,* Jeff Cole, "Star Wars: In New Space Race, Companies Are Seeking Dollars From Heaven," *Wall St. J.,* Oct. 10, 1995, p. A1, col. 6; John Markoff, "AT&T Plan Links Internet and Satellites, *N.Y. Times,* Oct. 4, 1995; Spectrum Sharing, Interference Concerns Dominate WRC-95 Comments to FCC," *Mobile Satellite News,* May 4, 1995.

[6]   47 C.F.R. §2.106; Adamson, *supra* Ch. 2, note 32, at 128, 130.

[7] *Id.* at 128–29.

[8] J. P. Schulz, "Little LEOs and Their Launchers," Vol. 3, *CommLaw Conspectus,* 1995, pp. 185–86.

[9] Pelton, *supra* note 3, at 51.

[10] *Id.*

[11] Notice of Proposed Rulemaking, In the Matter of Amendment of the Commission's Rules to Establish Rules and Policies Pertaining to Mobile Satellite Service in the 1610–1626.5/2483.5–2500 MHz Frequency Bands, 9 F.C.C. Rcd 1094 n. 6 (1994). The FCC has also referred to satellites in all of these orbits as "non-geostationary."

[12] Notice of Proposed Rulemaking, In the Matter of Amendment of Section 2.106 of the Commission's Rules to Allocate Spectrum to the Fixed-Satellite Service and the Mobile Satellite Service for Low-Earth Orbit Satellites, 6 F.C.C. Rcd 5932, n. 1 (1991) [hereinafter "Little LEO NPRM"].

[13] Schulz, *supra* note 8, at 186–87.

[14] "AT&T Plan Ambitious; At Least 52 Ka-Band Satellites Proposed in FCC Applications," *Communications Daily,* Oct. 5, 1995, p. 3. The proposal referenced is the Teledesic system, discussed in more detail in Ch. 18.

[15] Adamson, *supra* Ch. 2, note 32, at 132.

[16] Little LEO NPRM, *supra* note 12, at ¶ 9.

[17] Adamson, *supra* Ch. 2, note 32, at 132.

[18] 49 U.S.C. §§ 70101-70119 (1994).

[19] Schulz, *supra* note 8, at 189–90. The replacement for the satellite destroyed in the 1986 *Challenger* disaster was successfully launched by the space shuttle *Discovery* in 1995. *Communications Daily,* July 14, 1995, p. 11.

[20] Michael K. Frisby and Helene Cooper, "Clinton May Let Ukraine Launch U.S. Satellites," *Wall St. J.,* Dec. 6, 1995, p. A4, col. 4; Jeff Cole, "Re-Entry Mission: U.S. Rocket Makers Rely on Foreign Rivals for New Shot at Space," *Wall St. J.,* Nov. 10, 1995, p. A1, col. 6.

[21] *Communications Daily,* July 11, 1995, p. 6; "Cryogenic Upper Stage Added; Hughes Commits to 10 Launches on New MD Delta 3 Rocket," *Communications Daily,* May 11, 1995, p. 7.

[22] Martyn Williams, "Crash Impact: fallout from the recent spate of satellite crashes far and wide," *Multichannel News,* March 6, 1995, p. 22A.

[23] Schulz, *supra* note 8, at 190; Jeff Cole, "Star Wars: In New Space Race, Companies Are Seeking Dollars From Heaven," *Wall St. J.,* Oct. 10, 1995, p. A1, col. 6.

[24] *Id.; Communications Daily,* Aug. 8, 1995, p. 7; "Vitasat-1/Gemstar-1 Lost; Maiden Launch of Lockheed Vehicle Ends In $17-Million Failure," *Communications Daily,* Aug. 17, 1995; Martyn Williams, "Crash Impact: fallout from the recent spate of satellite crashes far and wide," *Multichannel News,* March 6, 1995, p. 22A (quoting PanAmSat President Fred Landman).

[25] The survey referred to showed the following rates, generally based on the preceding 30 launches of each: Delta 100%; Proton 96.7% (based on noncommercial launches); Ariane 93.3%; Long March 90%; Atlas 83.3%. Williams, *supra* note 24.

[26] Prospectus, American Mobile Satellite Corporation, at 6, Dec. 13, 1993.

[27] *Communications Daily,* May 10, 1995, p. 12.

[28] 47 C.F.R. § 2.1(c) (West 1995).

[29] John Davidson Thomas, "International Aspects of the Mobile Satellite Services," 43 *Fed. Com L.J.,* pp. 45, 47 (1990).

[30] Schulz, *supra* note 8, at 186 & n. 13.

[31] "Wireless Technology Reaches Out to Developing Economies," *Fin'l Post,* March 1, 1995; "Developing World: Globalstar Goes Low-End Route," *Dow Jones Int'l News,* Feb. 28, 1995.

[32] Little LEO NPRM, *supra* note 12, at ¶ 11; Pelton, *supra* note 3, at 36–37.

[33] Report, In the Matter of Preparation for International Telecommunication Union World Radiocommunications Conferences, 1995 FCC Lexis 3988, n. 12, 1995; Second Report and Order, In the Matter of Amendment of Parts 2, 22 and 25 of the Commission's Rules to Allocate Spectrum

for, and to Establish Other Rules and Policies Pertaining to the Use of Radio Frequencies in a Land Mobile Satellite Service for the Provision of Various Common Carrier Services, 2 F.C.C. Rcd 485, at ¶ 30, 62 R.R.2d 48 (1987).

[34]   *E.g.,* Memorandum Opinion and Order, In the Matter of AMSC Subsidiary Corporation Applications to Modify Space Station Authorizations in the Mobile Satellite Service, 8 F.C.C. Rcd 4040 at ¶ 3 (1993).

[35]   R. White, *supra* note 1, at 140.

[36]   47 C.F.R. § 25.103(d), (e) (1994).

[37]   Pelton, *supra* note 3, at 47–48.

[38]   Pamela L. Meredith and George S. Robinson, *Space Law: A Case Study for the Practitioner*, Boston: Martinus Nijhoff Publishers, 1992, pp. 74–75 [hereinafter "Meredith and Robinson"].

[39]   *Id.* at 49.

[40]   Saunders, *supra* Ch. 2, note 1, at 40; 60 & nn. 11–12, 19; 263; 336; 348–53.

[41]   Third Notice of Proposed Rulemaking and Supplemental Tentative Decision, In the Matter of Rulemaking to Amend Parts 1, 2, 21, and 25 of the Commission's Rules to Redesignate the 27.5–29.5 GHz Frequency Band, to Reallocate the 29.5–30 GHz Frequency Band, To Establish Rules and Policies for Local Multipoint Distribution Service and For Fixed Satellite Service, 1995 WL 447411 at ¶¶ 23–24 (1995). For a discussion of Teledesic, see Ch. 18, *infra.*

CHAPTER 16

▼▼▼

GLOBAL REGULATION

16.1   THE BACKGROUND OF THE ITU

The International Telecommunication Union is the ultimate successor to the International Telegraph Union, founded in 1865. It is an international body established by treaty (the International Telecommunication Convention or ITU Convention) [1] and is an agency of the United Nations. The ITU has global responsibility for the regulation of space communication [2].

The ITU (as opposed to its predecessor organizations) was established by the 1932 International Telecommunication Convention. The 1947 Telecommunication Convention made the ITU part of the UN system.

The organizational structure of the ITU was substantially modified in 1992. However, because pre-1992 source materials use the older terms, it is helpful to be aware of them as well as the current nomenclature.

Prior to this reorganization the periodic organs, which are called into session at intervals according to procedures set forth in the ITU Convention, were the Plenipotentiary Conference, Administrative Conferences, and the Administrative Council. In 1992, the Administrative Council was renamed the Council. The word "Administrative" was also dropped from the phrase "Administrative Conferences."

The pre-1992 Permanent Organs comprised the General Secretariat, the International Frequency Registration Board (IFRB), the Telecommunication Development Bureau, the International Radio Consultative Committee (CCIR), and the International Telegraph and Telephone Consultative Committee (CCITT). As a result of the 1992 reorganization, the functions of the IFRB were divided in two and

---

split between the Radio Regulations Board and the Radiocommunication Bureau. The duties of the CCIR and CCITT were assigned to the Radiocommunications Sector and the Telecommunication Standardization Sector. The Telecommunication Development Bureau became the Telecommunication Development Sector.

The Plenipotentiary Conference is the supreme organ of the ITU, with the power to revise the International Telecommunication Convention. It meets every four years. The Council is a sort of executive committee of the Plenipotentiary Conference and meets yearly in Geneva, the seat of the ITU. Radiocommunication Conferences (as opposed to Plenipotentiary Conferences) can be either world (WRC) or regional (RRC). These terms have since 1992 superseded the titles World and Regional Administrative Radiocommunication Conference (WARC and RARC). The ITU designates three regions of the world: Region 1 (Europe, Africa, the Middle East, the former USSR, and Mongolia); Region 2 (the Americas, the Caribbean, and Greenland); and Region 3 (Asia and the Pacific Basin other than Hawaii). WRCs may be either general or specialized. A Conference may revise the Radio Regulations that are annexed to the Convention. The Radio Regulations set forth the procedures to be followed by countries in coordinating the radio frequencies and orbital positions used by their satellites.

The four areas in which the duties of the General Secretariat fall are administrative, staff, support work for conferences and meetings, and information distribution. The Radio Regulations Board (and previously the IFRB) is charged with interpretation of the Radio Regulations and the application of those regulations to specific uses of radio frequencies that are notified to the ITU with requests for recordation in the Master International Frequency Register. The Radiocommunications Bureau has assumed the IFRB's secretarial functions: processing information received from national administrations; keeping records concerning frequency assignments and orbital positions and updating the Master International Frequency Register; and investigating complaints concerning harmful interference. The Radiocommunications Sector and the Telecommunication Standardization Sector, successors to the CCIR and CCITT, study telecommunications innovations and the creation of common international technical standards for radio communications and for wire communications, respectively. Based on their studies, they make recommendations to other organs or to members of the ITU.

Countries may make exceptions to either the ITU Convention or to the Radio Regulations. However, they may not then enforce the provision to which they have excepted against other members of the ITU.

The regulation of satellite communications has three important components: allocation, allotment, and assignment [3]:

> *Allocations* designate a group of frequencies to a service with priority *within* services decided largely on a first-come, first-served basis. *Allotments,* conversely, designate specific frequencies or bands of frequencies for the use of one or more identified countries or geographic areas under

specified conditions, and therefore more closely resemble national assignments. An *assignment* is an authorization given by an administration for a radio station to use a particular frequency or frequency channel under specified conditions.

National and regional governmental agencies may also make allocations of frequencies to a service where those allocations will not interfere with the ITU's allocations [4]. An ITU member may record its intention to deviate from a particular allocation by making a reservation to the Radio Regulations and placing a footnote in the Table of Frequency Allocations, which specifies the lawful uses for all radio-frequency bands between 9 kHz and 400 GHz.

As noted, once the ITU has decided to allocate a group of frequencies to a service or group of services (e.g., to one service on a primary basis and another on a secondary basis), priority among the satellite systems of the various countries within the service is largely on a first-come, first-served basis. The process, as more fully discussed in the next section, is referred to as *publication, coordination, and notification*. This system works as well as it does largely because of the nature of radio services. Transmission and reception equipment is essentially useless in the presence of harmful interference from other equipment. If a country were to launch a satellite in disregard of a previous registration made by another country, neither country's satellite would function properly. Moreover, a country that engaged in such behavior could not expect protection when the functions of its satellites were impinged upon by subsequent interfering countries.

This is not to say that every country has been happy with this *a posteriori* system of spectrum use. Spectrum planning, or *a priori* rights vesting, has been urged as an alternative. The next section provides an example of such dissatisfaction with the *a posteriori* method and a description of how it has been dealt with within the ITU.

16.2   THE ROLE OF DEVELOPING COUNTRIES IN ALLOCATION AND ALLOTMENT OF THE GSO

As noted, the GSO is a circular orbit above the equator at an altitude of 22,300 miles or 35,800 kilometers. By virtue of this definition, then, the GSO is a limited resource. Moreover, satellites in the GSO must be separated by a minimum amount of space to avoid radio interference and, what is less likely, physical collisions. As one might expect, the developed countries have a much larger number of satellites in the GSO than do the developing countries. Many developing countries have expressed concern that by the time they are ready to launch their own satellites, there will be no room in the GSO for them. Accordingly, they have sought to have part of the GSO reserved for them so that it will be available when they are prepared to use it. Because voting within the ITU has been on a one-vote-per-member basis since 1947 and the number of developing countries gaining independence has vastly increased over recent decades, the issue has been a persistent one [5].

The principle of *a posteriori* vesting of rights was established by the 1906 Convention of the International Radiotelegraph Union (which was later merged with the International Telegraph Union), although obviously not with respect to satellite communications. This process entailed classification of radio services (i.e., specifying frequency bands that are allocated to specific radio uses) and subsequent notification of national radio station assignments to a central, international recording agency. Thus, the international organization did not grant each user prior permission to use the spectrum, but recognized the right to use after the usage commenced.

A basic objective of the World Administrative Radio Conference, held in Atlantic City in 1947, was the establishment of a fully engineered (i.e., planned by negotiation) radio spectrum. However, the concept had to be dropped because, among other things, the requests submitted by ITU members far exceeded available frequencies. By 1959 it became apparent that the *a priori* method could not be used in all circumstances for use of the radio spectrum. A mixed approach was adopted. Rights to use would still be secured primarily by the *a posteriori* vesting method that proceeded on a first-come, first-served, case-by-case basis, but the highest degree of recognition and protection would be obtained through planning conferences where the spectrum was planned on an *a priori* basis.

Sputnik, the first artificial satellite, was launched in 1957. As a consequence, the ITU first allocated radio spectrum to space use in 1959.

Then, in 1963, the *a posteriori* principles were applied to space communications and the ITU's international coordination procedures were expanded. Countries were required to advise the IFRB of their intent to establish an international satellite system early in the planning process. The IFRB would publish this information, and coordination by the establishing country with other countries for the purpose of avoiding radio interference would occur. Then the IFRB would be notified and would record the assignment in the International Frequency Register. This three-step procedure for approval of orbital location and use of spectrum, which is still in place, is referred to as the publication, coordination, and notification process [6]. The system does not work perfectly: there are problems with countries that do not comply with the coordination procedures and others that advise the ITU of satellite systems that are never built [7].

Another important principle that was established in 1963 was that access to space should be on an equitable basis. This provided the developing countries with further support in their efforts to have spectrum reserved for them.

In the face of *a posteriori* rights vesting, the developing countries pursued several arguments. They have argued that registration of a frequency assignment should not give a permanent priority. A resolution to this effect was adopted in 1971, although it did not possess binding force. Another stance, this one taken since 1976 by some of the developing countries that are located in equatorial regions, has been to assert territorial claims to the segment of the GSO that is superjacent to their borders. However, this position is counter to the principles of the 1967 Outer Space Treaty [8].

Nevertheless, the developing countries did not fail to make progress. *A priori* schemes did to some degree displace *a posteriori* rights vesting with respect to the GSO. For example, in 1977, planning supplanted the first-come, first-served, case-by-case method with respect to certain broadcasting services in Regions 1 and 3 (i.e., the part of the world that does not include the Americas).

In 1985, the idea of a mixed approach was adopted specifically to ensure that every country would have access to at least a part of the GSO. This principle was implemented in 1988 with the establishment of an allotment plan. The plan has been described as follows [9]:

> The Allotment Plan was established in two parts. [Footnote omitted.] Part A provides allotments for all countries. Each country has a "nominal" orbital location that has a predetermined arc associated with it. The standardized parameters adopted are a compromise between existing technology and the somewhat "low tech" (i.e., less costly) standards preferred by developing countries. The generalized parameters adopted will permit implementation of most allotments without further technical coordination. By allotting each country an orbital position, part A satisfies the developing countries' demand for guaranteed access to the GSO.
>
> Part B of the Allotment Plan covers existing systems. [Footnote omitted.] These systems have a status equal to that of the national allotments in part A. At the conference, many administrations modified characteristics of their existing systems so that part B would be compatible, in most respects, with part A. By protecting existing systems, part B satisfies the developed countries' concerns about satellite systems using the Allotment Plan bands that are already deployed or are approaching implementation.

Thus, at least for the time being, the needs of developing countries for access to the GSO appear to have been satisfied.

16.3 THE ITU AND MSS

Another frequency allocation issue that has at times divided ITU members (although not on developing versus developed country lines) has been the allocation of spectrum to satellites that provide mobile services. Prior to 1987 there was no ITU allocation for land (as opposed to maritime and aeronautical) mobile services.

At WARC-MOB-87, the U.S. proposed that the ITU adopt allocations for MSS to be provided by GEOs that would parallel those already adopted by the United States. Several European countries opposed this, for two reasons. First, they asserted that such an allocation was unnecessary because of the lack of demand for LMSS beyond what could be satisfied by existing terrestrial systems (most of which were government run). Second, they contended that an allocation to land mobile would interfere with marine mobile satellite services being provided by INMARSAT [10]. Eventually, a compromise was worked out and some allocation was made. However, the United States took a reservation that stipulated its intent to continue its domestic allocation, which was much larger [11].

In 1992, the United States proposed additional allocations for MSS, this time including Big and Little LEOs. The U.S. proposal was accepted and new allocations were made [12]. The World Administrative Radio Conference (WARC-92) allocated the 1,610–1,626.5 MHz band for uplink transmissions and the 2,483.5–2,500 MHz band for downlink transmissions for Big LEOs. Shortly thereafter the FCC proposed an identical allocation and adopted that allocation in December 1993 [13]. It left open the possibility that the allocation could be used by GEOs as well as LEOs [14].

MSS allocations in the 2-GHz range have also been the source of some disagreement. The ITU allocated spectrum in this range to MSS in 1992. However, the United States then allocated some of that same spectrum to PCS [15]. The ITU decided to allocate additional 2-GHz frequencies to mobile satellites in 2005, but the United Kingdom sought to advance the effective allocation to the year 2000. Although this was opposed by countries that have terrestrial operations utilizing these frequencies [16], the ITU at WRC-95 in November 1995 did shift the allocation for regions other than North and South America to the year 2000 [17].

Notes

[1]   The current Constitution and Convention of the ITU were put forward at the Geneva Additional Plenipotentiary Conference which concluded on Dec. 22, 1992. Stewart White, Stephen Bate, and Timothy Johnson, *Satellite Communications in Europe: Law and Regulation*, Longman, 1994, p. 37 [hereinafter "S. White"].

[2]   The description of the ITU in this section is drawn largely from R. White, *supra* Ch. 15, note 1, particularly pages 29–105, and S. White, *supra* note 1, at Ch. 2.

[3]   R. White, *supra* Ch. 15, note 1, at 90.

[4]   Another important function of the ITU is the rendering of development assistance to developing countries. *Id.* at 66, 103, 115–16, 130–31, and 155–56.

[5]   The discussion in this section is drawn primarily from R. White, *supra* Ch. 15, note 1, at pp. 10–16, 23–24, 36, 54, 92–98, 114–30, 148–49, 172–73, 181–92, 201–18. The one-vote-per-member principle is set forth in Article 2 of the ITU Convention.

[6]   Tentative Decision, In the Matter of Amendment of Parts 2, 22 and 25 of the Commission's Rules to Allocate Spectrum for and to Establish Other Rules and Policies Pertaining to the Mobile Satellite Service for the Provision of Various Common Carrier Services, 6 F.C.C. Rcd 4900, at ¶¶ 37–41, 69 R.R.2d 828 (1991).

[7]   "Fines Off Limits; ITU Tries to Solve New Problems Without Interfering with National Sovereignty," *Communications Daily*, May 30, 1995, p. 4; "Response to ITU Resolution; U.S. Government and Industry Decide What to do About 'Paper Satellites,'" *Communications Daily*, Jan. 9, 1995, p. 5.

[8] Gregory C. Staple, "Current Development: The New World Satellite Order: A Report from Geneva," *A.J.I.L.*, Vol. 80, 1986, pp. 699, 711–12.

[9] Milton L. Smith, "Current Development: The Space WARC Concludes," *A.J.I.L.*, Vol. 83, 1989, pp. 596, 597–98.

[10] For a description of INMARSAT, see Ch. 18, *infra*.

[11] Thomas, *supra* Ch. 15, note 29, at 55–61. The ITU allocated only 7 MHz whereas the U.S. had sought 14, and the FCC had domestically allocated 27 MHz.

[12] Ted Stevens, "Comment, Regulation and Licensing of Low-Earth Orbit Satellites," *10 Computer and High Tech. L.J.*, 1994, pp. 401, 406; Regina A. LaCroix, "Developments in International Satellite Communications in the International Space Year," *CommLaw Conspectus*, Vol. 1, 1993, pp. 99, 105.

[13] Report and Order, In the Matter of Amendment of the Commission's Rules to Establish Rules and Policies Pertaining to a Mobile Satellite Service in the 1610–1626.5/2483.5–2500 MHz Frequency Bands, 9 F.C.C. Rcd 5936 at ¶ 8 (1994) [hereinafter "Big LEO Order"].

[14] The allocation was made in Report and Order, In the Matter of Amendment of Section 2.106 of the Commission's Rules to Allocate the 1610–1626.5 MHz and the 2483.5–2500 MHz Bands for Use By the Mobile-Satellite Service, Including Non-Geostationary Satellites, 9 F.C.C. Rcd 526 (1993). The FCC clarified that it did not find whether both GEOs and LEOs would be authorized in Memorandum Opinion and Order, In the Matter of Amendment of Section 2.106 of the Commission's Rules to Allocate the 1610–1626.5 MHz and the 2483.5–2500 MHz Bands for Use By the Mobile-Satellite Service, Including Non-Geostationary Satellites, 10 F.C.C. Rcd 3169 (1995). The test for when a GEO system might be permitted to use this spectrum was set out in Memorandum Opinion and Order, In the Matter of the Commission's Rules to Establish Rules and Policies Pertaining to a Mobile Satellite Service in the 1610–1626.5/2483.5–2500 MHz Frequency Band, FCC 96-54 at ¶ 31 (Released February 15, 1996).

[15] Notice of Proposed Rulemaking, In the Matter of Amendment of Section 2.106 of the Commission's Rules to Allocate Spectrum at 2 GHz for Use by the Mobile-Satellite Service, 10 F.C.C. Rcd 3230 (1995).

[16] "Hughes Gets Contract for INMARSAT Affiliate Satellites, May Invest in Project," *Mobile Satellite Rep'ts*, July 31, 1995; "Special Interview: U.S. WARC-95 Delegation Head Mike Synar and Staff, "*Mobile Satellite News*, July 13, 1995; "Crockett: Creation of INTELSAT Affiliate May be Politically Achievable," *Satellite Week*, April 17, 1995; "International Telecommunications Union Reserves 2 GHz Frequency Band for Proposed Global Satellite Network," *Computergram Int'l*, Feb. 28, 1995.

[17] "Satellite and International," *Communications Daily*, Nov. 20, 1995; "WRC-95 Decides Favorably On Several U.S. Proposals," *Mobile Satellite Rep'ts*, Nov. 20, 1995.

CHAPTER 17
▼▼▼

NATIONAL AND REGIONAL REGULATION

17.1 INTRODUCTION

As discussed in the preceding chapter, the ITU allocates spectrum, allots spectrum and orbital positions, and maintains a register of orbital positions and radio frequency assignments. Although the discussion of publication, coordination, and notification was primarily in the context of satellites in the GSO, the procedure applies as well to other types of satellites to which spectrum has been allocated.

However, the ITU Convention "fully recogniz[es] the sovereign right of each country to regulate its telecommunication,..." [1]. Thus, the contracting states retain those powers that they have not ceded to the ITU. Moreover, only a government, not a private satellite operator, may utilize the ITU's process for obtaining frequency and orbital position protection.

This leaves a substantial role for national governments and the regional administrations with which they choose to associate themselves, such as the European Union. Among the most important tasks that countries have retained the authority to perform are assigning orbital slots, licensing the provision of services over radio frequencies to persons physically located in a particular country, and approving the launch of satellites [2]. This chapter will discuss how some national and regional government entities accomplish these functions [3].

However, national and regional regulation is just as notable for what it fails to address as for what it does cover. Specifically, the law imposes little in the way of universal service obligations.

17.2 THE UNITED STATES

17.2.1 Authorization to Place in Orbit and Operate a Satellite

Before a satellite can be constructed, launched, and operated, its owner must first apply for and receive from the FCC a construction permit, a launch authorization, and an operating license. (The FCC has, however, proposed eliminating the construction permit requirement and allowing applicants to construct at their own risk while a decision on their operating license is pending [4].) The FCC launch authorization does not confer any right to launch or operate a launch vehicle, but rather allows for the satellite to be launched on a U.S. or foreign launch vehicle of its choice, subject to other restrictions. To launch a satellite on a launch vehicle operated by a private U.S. company, that company must apply for and receive a launch license from the Department of Transportation [5].

How one goes about obtaining a construction permit, launch authorization, and operating license from the FCC depends on (1) whether the satellite service to be provided is one which the FCC has already established, and (2) whether the service is to be provided using frequencies already allocated to it. If these two conditions are met, the filing of an application is appropriate. If not, then a petition for rulemaking is required to establish the service or to have the frequency allocated to it [6].

The FCC has authorized five categories of private commercial satellite operations, three of which will be covered here: domestic fixed satellite service, international fixed satellite service (*separate systems*), and mobile satellite service. (The other two are direct broadcasting service and radiodetermination satellite service [7].) As also discussed in this section, the FCC recently eliminated the distinction between domestic and international systems.

Before describing how each of these services were established and how they are regulated, a few preliminary words on the FCC's policies are in order. Generally, the FCC has had what has been referred to as an *open entry, multiple entry,* or *open skies* policy when it comes to satellite licensing. The FCC has never used comparative hearings as a way to determine which systems should be approved. Since this can lead to having more applications than there are frequencies and orbital positions, some way had to be found of limiting the authorizations awarded. The FCC has accomplished such limitations by, where necessary, imposing technical, financial, and legal requirements that have disqualified some applicants [8]. As will be discussed in this chapter, the question that has arisen is how to resolve problems of mutual exclusivity when these hurdles are not enough. One possible answer that has been considered by the FCC is spectrum auctions, as have been used with PCS.

All applicants for space station authorizations must submit comprehensive proposals. Among the data to be submitted are the identity of the applicant and its legal counsel; the type of authorization requested; a description of the system; technical information about the system (such as frequencies used, orbital position sought, and detailed technological aspects of the system); estimated number and distribution of Earth stations; capabilities to provide service; historical use of the system if it already exists (for applications for additional or replacement satellites); arrangements for tracking, telemetry and control; physical characteristics of the space station(s); schedule of annual costs and revenues for the life of the system as well as other financial information; whether the system is to be operated on a common carrier basis; schedule for construction and launch; public interest considerations in support of the grant; and any other information specifically applicable to the particular type of satellite service to be provided [9].

One question that recurs in satellite regulation is whether satellite operators will be treated as common carriers. A common carrier must offer its services to the public on reasonable and nondiscriminatory terms and may be subject to tariffing obligations. Private carriers, on the other hand, may offer service to selected customers on negotiated terms. Common carriers may be considered either dominant or nondominant, depending on whether they possess market power, and their rate setting and tariffing obligations vary according to their classification [10].

The FCC licensed the first private commercial U.S. domestic fixed satellites (domsats) in 1973, when five companies were authorized to construct systems that would provide fixed satellite services solely within the United States [11]. U.S. licensing of domestic satellite service is of interest to developing countries for two reasons. First, it may provide useful information for developing countries interested in creating a licensing scheme for their own domestic or regional satellite systems. Second, U.S. domsats have been permitted, pursuant to an FCC doctrine called the *transborder policy*, to serve countries other than the United States if those other areas fall within their footprint.

The first application by a private company for authority to operate a domestic satellite system was filed in 1965. The FCC had to decide whether it had the legal authority to grant such licenses, whether the public interest would be served by such grants, and how the grants should be made. The FCC first concluded, in 1970, that it could authorize any nonfederal governmental entity to construct and operate satellites for domestic purposes and that while it could not decide which systems to authorize based on the record before it, it would accept applications from all legally, technically, and financially qualified entities. It allowed applicants to "propose the rendition of such services to the public directly on a common-carrier basis or by lease of facilities to other common carriers, or any combination of such arrangements" [12]. The policy of multiple or open (but not unrestricted) entry was adopted two years later. The FCC did not want to risk depriving the public of the benefits of satellite technology by requiring new entrants to show that competition is reasonably feasible and that the anticipated market can economically support the proposed facilities [13]. (Essentially, the FCC adopted what might now be characterized as the "build it and they will come" approach.) The FCC subsequently began granting domsat construction permits and licenses in 1973 [14].

By the 1980s, the FCC had granted enough satellite authorizations under the open entry approach to realize that it would have to modify its policies if it wished to accommodate more satellites. It reduced the amount of spacing required between satellites in the GSO from four degrees in the C band and three degrees in the Ku band to two degrees in both [15].

The next step taken by the FCC so that it could maintain its open entry policy was to impose stringent financial qualifications that resulted in elimination of sufficient applicants that those remaining could be accommodated. The justification advanced for such qualifications was that they were needed to avoid tying up valuable spectrum and orbital positions [16]. The FCC requires applicants for domestic fixed satellite space station licenses to demonstrate their "current financial ability to meet the (1) estimated costs of proposed construction and/or launch, and any other initial expenses for the space station(s); and (2) estimated operating expenses for one year after launch of the proposed space stations." Detailed financial information must be submitted to support this showing" [17].

Domsats were initially considered common carriers with respect to both sales and leases of transponders [18]. They were also at first considered to be dominant carriers. However, they were subsequently classified as nondominant and subject to streamlined tariffing obligations. Moreover, since the initial licensing of domsats, the FCC has allowed domsat operators to sell transponders [19] on a non-common-carrier basis. At first this exemption for sales was conditioned upon a finding that granting a particular sales request would not unduly reduce the number of transponders available on a common-carrier basis. [20] However, the FCC then abandoned case-by-case scrutiny [21]. Ultimately, the FCC decided that there was no longer a need to require domsat licensees to provide capacity on a common-carrier basis and that it would allow satellite operators to elect to operate on a common carrier or non-common-carrier basis [22].

Although satellites in the fixed domestic service were intended primarily to serve the United States, the FCC instituted a transborder exception to allow some foreign service to be provided. In 1981, the State Department concluded that under some circumstances such an exception would not be inconsistent with the United States' obligations under the International Telecommunication Satellite (INTELSAT) Agreement. INTELSAT, discussed in Chapter 18, is an international organization under which many of the world's countries have agreed to operate a global satellite system. The INTELSAT Agreement limits participants' ability to operate or authorize the operation of other international satellite systems that might compete with INTELSAT. The State Department's 1981 opinion, referred to as the *Buckley Letter*, was the basis for the Transborder Policy, under which U.S. domsats were authorized to provide international services where INTELSAT could not provide the service or where it would be clearly uneconomical or impractical to use INTELSAT facilities.

(The Buckley letter criteria were revised to apply only where more than 1,250 circuits are to be provided under the Transborder Policy.) Although service to the Caribbean and Central and South America has been authorized under the Transborder Policy, it has generally been easier to obtain permission to serve Canada and Mexico, which are contiguous to the United States [23].

The United States is limited by the INTELSAT Agreement in its ability to authorize the operation of private international (as well as the previously discussed domestic) fixed satellite systems. The first application for such a "separate system" was filed in 1983. In 1984, the President determined that separate systems are required in the national interest. At the same time, the U.S. Secretaries of State and Commerce jointly advised the FCC to authorize separate systems provided that [24]:

(1) each system be restricted to providing services through the sale or long-term lease of capacity for communications *not interconnected with public switched message networks* (except for emergency restoration service); [Footnote omitted] and (2) each system gain approval from the foreign authority with which communications links are being established and enter into consultation in accordance with Article XIV(d) of the INTELSAT Agreement to ensure technical compatibility and to avoid significant economic harm to INTELSAT.

The absolute prohibition on interconnection with the PSTN was instituted to avoid competition with INTELSAT's core business. However, it has been relaxed with the passage of time. Over the years, INTELSAT has presumed that a particular number of circuits of a separate system can be allowed to interconnect without a showing of no economic harm to INTELSAT. That number has been increasing and the limit is to be abolished entirely as of January 1, 1997. Separate systems may also provide private line services interconnected with the PSTN, which permit private line customers to use a single system of customer premises equipment (both telephone and computer) to access a mix of switched and nonswitched telecommunications services. Moreover, separate system licensees may provide domestic service within the United States on an "ancillary" basis [25].

Like domsat applicants, prospective operators of separate systems were required to demonstrate the financial ability to assume the costs and liabilities involved in constructing, launching, and—for one year—operating the system. However, the FCC adopted a two-step approach to determining the financial qualifications of separate systems applicants. A separate system must "consult" with INTELSAT on technical and economic issues, and it is unlikely to receive irrevocable financial commitments until the consultation process is completed. In the first stage, the applicant was required to show "(1) the estimated costs of proposed construction and launch, and any other initial expenses for the proposed space station(s); (2) the estimated operating expenses for one year after launch of the proposed space station(s); and (3) the source(s) or potential source(s) of funding of the proposed system for one year, which would include the identity of financiers and their letters of financial interest." A conditional construction permit would be issued upon such a showing, not to permit construction to commence (although a waiver may be obtained that will allow an early start on construction [26]), but to set forth the technical parameters needed for INTELSAT consultation. In the second stage, an order permitting construction will be issued upon a showing of current financial ability to meet the costs of construction, launch, and operating expenses for one year after launch [27].

Space segment operators of separate systems are not classified as common carriers. However, common carriers may resell separate system satellite capacity [28].

The FCC has decided to abolish the distinction between its Transborder Policy and its Separate International Systems Policy. Since those policies were instituted in the 1980s, the FCC has observed an increasing trend toward globalization of the economy. Users of communications systems, particularly multinational corporations, have greater need of international communications and would like to be able to procure such services from one vendor. However, separate licensing of domestic and international satellite systems can prevent a provider licensed as one from providing the other type of service.

The FCC therefore decided to abolish the Transborder Policy (including the Buckley letter criteria) entirely and to treat domsat licensees on the same basis as separate systems, allowing all U.S.-licensed fixed satellite systems to provide domestic and international services. Concomitantly, the restriction on separate systems that they only be permitted to provide domestic service on an ancillary basis has been abolished. Differences in financial qualification requirements for domsats and separate systems have been reconciled by eliminating the two-stage process for most separate systems [29].

The third type of private, commercial satellite service to be discussed here (after domestic fixed and international fixed) is the mobile satellite service. The analysis will focus on land (as opposed to aeronautical and maritime) mobile services, and particularly on the two-way voice services that could substitute for the basic landline telephone system in developing countries.

The FCC first allocated spectrum to MSS in 1986, pursuant to a notice of proposed rulemaking that it had adopted in 1984. The first license to provide land mobile satellite service was issued to a consortium of applicants, known as the American Mobile Satellite Corporation (AMSC). The FCC imposed the consortium requirement when a number of applicants sought the spectrum. Rather than conduct comparative hearings, which it felt would cause a delay that could jeopardize the U.S. position in international negotiations to obtain spectrum for MSS, it took the consortium approach. It established a financial test for applicants: to qualify, the interested applicants had to deposit five million dollars into an escrow account. (This requirement was later reduced for certain participants who challenged the minimum contribution requirement in court.) It further decided that AMSC would be regulated as a nondominant common carrier [30]. (AMSC is discussed in more detail in Chapter 18.)

The nonvoice, nongeostationary (NVNG) satellite services provided by Little LEOs (i.e., at frequencies below 1 GHz) were proposed to be established in 1991, after applications had been filed during 1990 [31]. In 1992, the FCC established a Federal Advisory Committee of affected parties to negotiate rules for Little LEOs. The Commission uses negotiated rulemaking in the hope that it will "achieve better, less contested regulations by involving interested parties in a face-to-face, pre-rulemaking process of cooperation and discussion" [32]. The Committee determined that there was adequate spectrum available to accommodate all the applicants, with room left over for future entry. Accordingly, no requirement that a Little LEO system provide domestic coverage for a specific period of time was imposed [33].

Financial qualifications for applicants seeking to provide NVNG satellite services are set forth in the FCC's regulations. Each applicant must show [34]:

> that it is financially qualified to proceed expeditiously with the construction, launch, and operation for one year of the first two space stations of its proposed system immediately upon grant of the requested authorization.

Milestone deadlines, the dates by which construction of a system must be commenced and completed and satellites launched, were not codified but are established in each individual license. Generally, however, a permittee must commence construction of the first two satellites of the system within one year of grant of the construction permit and begin construction of the remaining satellites within three years of grant. Construction of the first two satellites must be completed within four years of the grant and the entire system must be launched and operational within six years [35].

The FCC originally proposed that NVNG applicants be allowed to request classification as either common or private carriers. After that proposal was made, Congress amended the Communications Act to create the service category of commercial mobile radio service (CMRS) and required that CMRS providers be treated as common carriers, while allowing the FCC to determine whether provision of space segment capacity by satellite operators is CMRS. The FCC decided that provision of the space segment should not be treated as common carriage, unless the operator so chooses. Earth station licensees providing NVNG as CMRS would be treated as common carriers. However, any space or Earth segment licensee considered a common carrier would be subject to streamlined regulation [36].

The first applications for Big LEOs (voice and data services at frequencies above 1 GHz) were filed in late 1990. They were placed on public notice for comment and additional applications to be considered concurrently with them were filed by the deadline established by the FCC. The FCC then established an "MSS Above 1 GHz Negotiated Rulemaking Committee" in 1992. The Committee was in existence for three months to develop rules and policies. Its recommendations were the basis for rules proposed by the FCC in early 1994 [37].

In an October 1994 Report and Order, the FCC set out the licensing procedures for Big LEOs [38]. The FCC stressed from the outset its policy of limiting open entry by establishing qualification hurdles when necessary, given practical limitations [39]:

> [U]nless otherwise proscribed by rule, statute, or treaty, the Commission has traditionally adopted qualification requirements for each satellite service that reflect the nature of and entry opportunities for the particular service being licensed. Where entry opportunities for a particular service are limited, our threshold qualification requirements for that service are designed to ensure that those awarded licenses can expeditiously implement state-of-the-art systems that further the public interest. If applicants are unable to meet the basic qualifying criteria, their applications are dismissed without additional hearing.

With respect to Big LEOs, the FCC adopted global coverage requirements and strict financial qualifications and stated further that if these did not resolve problems of mutual exclusivity, it might implement a spectrum auction procedure.

Two coverage prerequisites were imposed. First, Big LEO systems must be capable of serving locations as far north as 70 degrees latitude and as far south as 55 degrees latitude. (The 70-degree north latitude line traverses the northernmost parts of Alaska and Norway. The 55-degree south latitude line is at approximately the southernmost tip of South America and runs below the continent of Africa and the countries of Australia and New Zealand.) These lines were drawn to cover most populated areas without imposing the cost of serving the few populated areas in or near the polar regions. The second condition is that each LEO system must have at least one satellite at an elevation angle of at least 5 degrees covering all 50 of the United States, Puerto Rico, and the U.S. Virgin Islands [40].

The FCC concluded that a strict financial requirement is warranted for Big LEO service. It noted that the proposed systems would cost between $97 million and $2 billion to implement and that financing is often difficult to obtain. It was concerned that if the standards were too lax, spectrum might be tied up for an ultimately unsuccessful planned system. Accordingly, it adopted a financial test identical to the one used for domsats: that applicants provide evidence of current assets, operating revenues, or irrevocably committed debt or equity financing sufficient to meet the estimated costs of constructing and launching all planned satellites, and operating the system for the first year [41].

The FCC also addressed the issue of regulatory treatment: whether Big LEO licensees would be treated as common carriers. To the extent that Big LEO space station licensees provide service directly to end users and this service meets the definition of commercial mobile radio service (CMRS), they will be treated as common carriers, just as are other CMRS providers (e.g., cellular, paging, PCS). However, if space segment capacity is offered to a reseller or other entity who then offers CMRS to end users, the FCC determined that there is no need at this time to impose common carrier obligations on the licensees, although they may elect common carrier status if they wish. In such a case, they would be considered nondominant. The FCC, in making this determination, relied primarily on the existence of competition. However, it also recognized that the global nature of LEO systems would require the raising of capital in the international markets and the willingness of foreign governments to allow a U.S. licensee to operate within their own borders. Radio common carriers are subject to limitations on foreign ownership under Section 310(b) of the Communications Act, which could interfere with plans to achieve global coverage [42].

A timetable for system implementation will be included in each license. Generally, each licensee will be required to begin construction of its first two satellites within one year of the unconditional grant of its authorization and complete construction of those first two satellites within four years of that grant. Construction for the remaining authorized operating satellites in the constellation must begin within three years of the initial authorization, and the entire authorized system must be operational within six years. A different schedule may be allowed if the applicant can show that the size or complexity of the system justifies additional time [43].

The FCC expressly declined to impose public service obligations on Big LEO licensees at that time. Although it was urged by some commenters that service be made available to educators and students at preferential rates, the FCC found that "there is not sufficient information in this record to support such requirements at this time" [44].

The FCC has proposed allocating spectrum in the Ka band to GEO and LEO fixed satellite service [45]. It has asked for comment on a proposal to apply the existing rules (including financial qualification standards) that it uses for other systems utilizing the geosynchronous orbit (e.g., domsats) to GEOs using the Ka band for FSS. It has further requested comment on what sort of rules should be created for LEOs using the Ka band for FSS. If all qualified applicants cannot be accommodated, the FCC proposes to use competitive bidding to award licenses to provide domestic service in the United States [46].

17.2.2 Landing Rights

The term *landing rights* refers to the right to transmit signals to and/or from the territory of a country other than the one that licensed the satellite system. Landing rights may be obtained by the satellite system operator or through a service agreement with a foreign telecommunications entity operating in the country [47]. Landing rights are of obvious importance for any satellite operator who wishes to serve more than one country or provide a global roaming service.

Thus, once a prospective U.S. satellite operator has received its FCC authorizations to operate a system, it must, unless its system is purely domestic, obtain landing rights from other countries. The procedure for receiving such permission varies from country to country and some administrations may not have a formal process [48].

Procuring investment by foreign telecommunications operators in U.S.-licensed satellite systems has been suggested as a way of acquiring landing rights for mobile satellite services [49]. INTELSAT consultation is also a prerequisite to obtaining foreign landing rights from countries that are parties to the INTELSAT treaty [50].

But how does a non-U.S.-licensed operator obtain landing rights in the United States? To operate as a carrier, it must obtain an authorization under Section 214 of the Communications Act [51]. A common-carrier applicant is subject to the foreign ownership restrictions of the FCC's rules. Any foreign carrier must also obtain Earth station licenses for its gateway stations and subscriber terminals [52].

Efforts by a non-U.S operators to obtain landing rights in the United States have sometimes met opposition. One of the more prominent examples arose out of the practice of the government of the small island nation of Tonga of filing at the ITU for orbital slots and frequencies and then leasing them to other countries or private operators for profit [53]. When Rimsat, one of the prospective users of slots for which Tonga had filed at the ITU, sought U.S. landing rights, its request was opposed on the ground that countries are not allowed, under ITU standards, to make requests for orbital locations and frequencies in excess of their own requirements [54]. The FCC has not yet ruled on the matter. In another case, the FCC refused to deny Section 214 applications to establish links between the U.S. PSTN and satellites owned, operated, or licensed by the Russian Federation, finding no merit to allegations by a party that opposed the applications that it had been denied landing rights in Russia [55].

17.2.3   Review of Satellite Licensing Policies

The FCC is currently in the process of reviewing many of the satellite licensing policies described above. In September 1995 the International Bureau of the FCC issued a public notice seeking industry dialogue. Among the questions being investigated are the following:

General Satellite Licensing Questions

Are the financial qualification standards applied by the Commission both necessary and sufficient to ensure that satellite applicants have the financial ability to implement their proposed system?

In the future, should the Commission license satellites in "rounds" (as we have for domestic satellites and other services), on a case-by-case basis (as we have for separate systems), or through some other method?

For geostationary satellites, should mutual exclusivity be determined orbital position by orbital position, or can the orbital arc be divided into segments within which each orbital position is fungible with each other position?

Fixed Satellite Service (FSS)

Are there technical or other developments that may make it possible to separate FSS satellites in geostationary orbit by less than two degrees?

How should the Commission license fixed Earth stations that communicate with nongeostationary satellite systems? What sort of information should the Commission require from applicants for such licenses?

Mobile Satellite Service (MSS)

What sort of license should the Commission issue to MSS systems whose satellites are already licensed by another nation, but who wish to operate in the United States? What sort of information should the Commission require from applicants for such a license? Should such licensees be treated as common carriers, as commercial mobile radio service providers, or under some other regulatory classification?

What licensing models are other countries using to regulate services by MSS systems whose satellites are licensed by the United States? How should administrations work to complement each other's licensing schemes?

What steps can the Commission take to facilitate international roaming for MSS systems?

What can be done to streamline the process for obtaining equipment approvals around the world, particularly for mobile Earth terminals?

Resolving Mutual Exclusivity

What unique policy questions do auctions present in the context of domestic and international satellite services?

If mutual exclusivity arises, which of the FCC's currently authorized selection mechanisms—auctions, lotteries, and comparative hearings—will best serve the public interest? Are there any combinations of these mechanisms—or rules governing these mechanisms—which would make them better serve the public interest?

Are there any selection mechanisms *other* than auctions, lotteries, and comparative hearings that should be considered?

The FCC hopes that the information it receives during the informal comment stage will form the basis for a formal rulemaking proposal beginning in 1996 [56].

17.3   THE EUROPEAN UNION

The regulatory situation in most of the countries of the European Union at the end of the 1980s stood in contrast to the open skies policies of the United States [57]. Accordingly, the EU saw a need to study and propose reforms with respect to satellite regulatory policy. (For a discussion of the lawmaking process in the EU, see Section 7.2.)

In November 1990, the European Commission issued a Green Paper on Satellite Communications [58]. The Satellite Green Paper recognized that regulation of the Earth and space segments in the Member States at the time reflected, in most cases, the situation in the 1960s and 1970s "where the only technically and economically feasible application of satellite communications was their use as an additional transmission path to carry international or national long-distance traffic for Telecommunications Organizations" [59]. (Telecommunications organizations, or TOs, are the national telephone companies, which have historically been government-owned monopolies.)

The Satellite Green Paper was intended to apply the EU's principles of telecommunications policy (liberalization and harmonization) to the satellite sector in order to encourage development of Communitywide systems and to control problems arising from competition for satellite capacity. In order that the full advantages of satellite communications services be realized, four major regulatory changes were proposed [60]:

- Full liberalization of the Earth segment, including both receive-only and transmit/receive terminals, subject to appropriate type-approval and licensing procedures where justified to implement necessary regulatory safeguards;
- Free (unrestricted) access to space segment capacity, subject to licensing procedures in order to safeguard those exclusive or special rights and regulatory provisions set up by Member States in conformity with Community law and based on the consensus achieved in Community telecommunications policy.

  Access should be on an equitable, non-discriminatory and cost-oriented basis.
- Full commercial freedom for space segment providers, including direct marketing of satellite capacity to service providers and users, subject to compliance with the licensing procedures mentioned above and in conformity with Community law, in particular competition rules;
- Harmonization measures as far as required to facilitate the provision of Europe-wide services. This concerns the mutual recognition of licensing and type approval procedures, frequency coordination and coordination with regard to Third Country providers.

However, the Satellite Green Paper recognized that exclusive or special rights may be entrusted to TOs for the provision of terrestrial public network infrastructure and public voice telephony service. The latter was defined as "the commercial provision for the public of direct transport of real-time speech via the public switched network or networks such that any user can use equipment connected to a network termination point to communicate with another user of equipment connected to another termination point" [61]. (Nevertheless, in separate developments the EU is also in the process of liberalizing basic telephone service, for which monopolies must be abolished, with a few exceptions, by 1998.)

The Satellite Green Paper was approved by the Council of Ministers in a resolution in December 1991 [62]. The Green Paper's general goals were endorsed by the Council. Implementing legislation was to be introduced by the Commission by the end of 1992 (a deadline that was not fully met) [63].

However, since the issuance and approval of the Satellite Green Paper, progress has been made, to varying degrees, on four fronts. These are: pan-European type approval of Earth station equipment; liberalization of satellite equipment and services; creation of EU-wide licensing; and development of satellite personal communication services.

The Commission published a further proposal, on December 10, 1992, for a Council Directive enabling Earth station equipment tested in one Member State to be freely available in another. The proposed directive also harmonizes requirements for obtaining approval of such equipment and guarantees the right to connect ground station equipment to telecommunications networks. The Directive on Pan-European Type Approval of Satellite Earth Station Equipment was adopted by the Council of Ministers in late 1993. However, it does not apply to Earth station equipment intended for use as part of a Member State's public telecommunications network [64].

The Commission has used its powers to promote competition to adopt a directive amending a 1988 EU law mandating competition in telephony generally to cover satellite equipment and nonbroadcasting services. Satellite communications had previously been excluded from this law. This directive, issued on October 13, 1994, requires elimination of all remaining exclusive rights for import, trade, connection, operation, and maintenance of satellite equipment and provision of satellite services, such as mobile satellite networks (like fleet tracking), VSAT networks, and specialized private networks. It does not apply to public-switched voice telephony. Moreover, states with insufficiently developed terrestrial networks may defer reporting on compliance until 1998. Some Member States (such as the UK, Germany, the Netherlands, and France) had already adopted liberalizing measures in these areas [65].

The next effort has been to create a scheme under which a license to operate a satellite network or provide satellite services issued in one Member State would be recognized in all others. The idea is that this would make it easier to establish EU-wide satellite networks. A draft proposal for a Council Directive on this subject was put forward by the Commission in October 1992, and the proposal was finalized in 1994. The proposal would require that in order to be entitled to mutual recognition, the license would only apply to firms that are at least 75% EC-owned. Again, some Member States already have bilateral agreements to accomplish this type of recognition [66].

The EU recognizes that the United States is further along in the process of licensing Big LEO systems to provide these services and is concerned that it might not have sufficient influence over how such systems are implemented. In December 1993, the Council of Ministers adopted a resolution (which, unlike a directive, does not require Member State compliance) inviting Member States to coordinate their individual actions and policies in connection with satellite personal communications services [67]. Most recently, in November 1995, the European Commission asked for the right to decide what companies should be licensed to operate these systems within the Community. The Commission said that it intends to establish categories of service by September 1996, adopt criteria for selection of systems by October 1996, select systems based on "comparative bidding" (possibly involving the use of auctions) by December 1996, and adopt harmonized conditions for authorization of services by March 1997. The next step for this proposal would be approval by the European Parliament [68].

Notes

[1] International Telecommunication Convention, *supra* Ch. 16, note 1, Preamble.
[2] Licensing Earth stations and permitting interconnection to the other communications networks (particularly the public-switched telephone network) are two others. Interconnection with the PSTN and EU Earth station licensing are touched on in this chapter. For coverage of United States law on the subject of licensing Earth stations, see, *e.g.*, Morton I. Hamburg and Stuart N. Brotman, *Communications Law and Practice* § 6.01[5], New York: Law Journal Seminars Press, 1995.
[3] For a description of satellite regulation in the United Kingdom, see S. White, *supra* Ch. 16, note 1, at Ch. 6.
[4] Notice of Proposed Rulemaking, In the Matter of Streamlining the Commission's Rules and Regulations for Satellite Application and Licensing Procedures, at ¶¶ 7–8, IB Docket No. 95-117 (1995).
[5] Meredith and Robinson, *supra* Ch. 15, note 38, at 115.
[6] *Id.* at 111, 114.
[7] *Id.* at 72. For definitions of the various types of satellite services, see 47 C.F.R. § 2.1(c).
[8] *E.g.*, Michael S. Straubel, "Telecommunications Satellites and Market Forces: How Should the Geostationary Orbit be Regulated by the FCC?," 17 *N.C. J. Int'l L. & Com. Reg.*, 1992, p. 205, text accompanying notes 60–92 (1992).
[9] 47 C.F.R. § 25.114 (1994).
[10] Kennedy, *supra* Ch. 4, note 12, at 85–92, 114–17. For a description of the FCC's current policy on classifying carriers as dominant or nondominant, and the ramifications thereof, see, *e.g.*, Order, In the Matter of Motion of AT&T Corp. to be Reclassified as a Non-Dominant Carrier, FCC 95-427 (Oct. 12, 1995).

[11] Notice of Proposed Rulemaking, In the Matter of Amendment to the Commission's Regulatory Policies Governing Domestic Fixed Satellites and Separate International Satellite Systems, 10 F.C.C. Rcd 7789 at ¶ 4 (1995).

[12] Report and Order, In the Matter of Establishment of Domestic Communication-Satellite Facilities by Nongovernmental Entities, 22 F.C.C. 2d 86 at ¶ 19 (1970).

[13] Second Report and Order, In the Matter of Establishment of Domestic Communication-Satellite Facilities by Nongovernmental Entities, 35 F.C.C. 2d 844 at ¶¶ 15–18 (1972), recon. in part, 38 F.C.C. 2d 665 (1972).

[14] Meredith and Robinson, *supra* Ch. 15, note 38, at 78–79.

[15] Memorandum Opinion and Order, Licensing of Space Stations in the Domestic Fixed Satellite Service and Related Revisions of Part 25 of the Rules and Regulations, 93 F.C.C.2d 1260 (1983).

[16] Meredith and Robinson, *supra* Ch. 15, note 38, at 81–82.

[17] 47 C.F.R. § 25.140(c) (1994).

[18] Kennedy, *supra* Ch. 4, note 12, at 116–17.

[19] The transponder is the active communications portion of the satellite. It receives the incoming signal, filters it, translates the incoming frequency to the outgoing frequency, amplifies it, and sends the signal forward to the antenna system for relay to the intended destination. Pelton, *supra* Ch. 15, note 3, at 261.

[20] Memorandum Opinion, Order and Authorization, Domestic Fixed-Satellite Transponder Sales, 90 F.C.C. 2d 1238, 1253-55 (1982), aff'd sub nom., Wold Communications, Inc. v. F.C.C., 735 F.2d 1465 (D.C. Cir. 1984), modified, Martin Marietta Communications Systems, 60 R.R.2d 779 (1986).

[21] Hamburg and Brotman, *supra* note 2, at p. 6–14.

[22] Report and Order, In the Matter of Amendment to the Commission's Regulatory Policies Governing Domestic Fixed Satellites and Separate International Satellite Systems, FCC 96-14 at ¶¶ 45–50 (Released January 22, 1996).

[23] *NPRM, supra* note 11, at ¶¶ 4–8, 14; Meredith and Robinson, *supra* Ch. 15, note 38, at p. 74, note 6.

[24] NPRM, *supra* note 11, at ¶ 10 (emphasis supplied). For discussions of the history of the licensing of separate systems in the United States, see Meredith and Robinson, *supra* Ch. 15, note 38, at 96–110; Straubel, *supra* note 8, at text accompanying notes 93–139.

[25] NPRM, *supra* note 11, at ¶¶ 11–14, 18.

[26] "FCC News Release: PanAmSat Granted Waiver to Spend Additional Funds to Construct Its International Separate Satellite (PAS-4)," *FCC Daily Digest*, July 20, 1993.

[27] Report and Order, In the Matter of Establishment of Satellite Systems Providing International Communications, 101 F.C.C. 2d 1046, at ¶¶ 233–36 (1985).

[28] *Id.* at pp. 1050, 1112.

[29] Report and Order, *supra* note 22; Notice of Proposed Rulemaking, *supra* note 11, at ¶¶ 1, 18–19, 21, 25, 29, 31–32.

[30] Notice of Proposed Rulemaking, In the Matter of Amendment of Parts 2, 22, and 25 of the Commission's Rules to Allocate Spectrum for, and Establish Other Rules and Policies Pertaining to, the use of Radio Frequencies in a Land Mobile Satellite Service for the Provision of Various Common Carrier Services, 50 Fed. Reg. 8149 (Feb. 28, 1985); Report and Order, In the Matter of Amendment of Parts 2, 22, and 25 of the Commission's Rules to Allocate Spectrum for, and Establish Other Rules and Policies Pertaining to, the use of Radio Frequencies in a Land Mobile Satellite Service for the Provision of Various Common Carrier Services, 2 F.C.C. Rcd 1825 (1986); Final Decision on Remand, In the Matter of Amendment of Parts 2, 22, and 25 of the Commission's Rules to Allocate Spectrum for, and Establish Other Rules and Policies Pertaining to, the use of Radio Frequencies in a Land Mobile Satellite Service for the Provision of Various Common Carrier Services, 7 F.C.C. Rcd 266, at ¶¶ 25–34 (1992).

[31] Little LEO NPRM, *supra* Ch. 15, note 12.

[32] Notice of Proposed Rulemaking, In the Matter of Amendment of the Commission's Rules to Establish Rules and Policies Pertaining to a Non-Voice, Non-Geostationary Mobile-Satellite Service, 8 F.C.C. Rcd 6330 at ¶ 4 (1993).

[33] *Id.* at ¶¶ 8–9; Report and Order, In the Matter of Amendment of the Commission's Rules to Establish Rules and Policies Pertaining to a Non-Voice, Non-Geostationary Mobile-Satellite Service, 8 F.C.C. Rcd 8450 at ¶¶ 22–23 (1993).

[34] 47 C.F.R. § 25.142(a)(4) (1994).

[35] Report and Order, *supra* note 33, at ¶ 18.

[36] *Id.* at ¶ 24.

[37] Notice of Proposed Rulemaking, In the Matter of Amendment of the Commission's Rules to Establish Rules and Policies Pertaining to a Mobile Satellite Service in the 1610–1626.5/2483.5–2500 MHz Frequency Bands, 9 F.C.C. Rcd 1094 (1994). One of the applications filed in this round was for a GEO, rather than a LEO system.

[38] Big LEO Order, *supra* Ch. 16, note 13.

[39] *Id.* at ¶ 11.

[40] *Id.* at ¶¶ 21–25.

[41] *Id.* at ¶¶ 26–42.

[42] *Id.* at ¶¶ 171–181. For a discussion of CMRS, see Ch. 9, *supra*. For a discussion of the foreign ownership restrictions of Section 310(b), see Section 4.6, *supra*.

[43] *Id.* at ¶ 189.

[44] *Id.* at ¶¶ 204–206.

[45] The FCC's plan is to segment the spectrum from 27.5–30 GHz, which the FSS will share with local multipoint distribution services (also referred to as cellular video or wireless cable TV) and feeder links for MSS. The 29.5-30 GHz band is currently allocated to MSS, although there are no systems operating in it. The FCC has proposed to eliminate this allocation or to modify it to make it secondary to FSS. Third Notice of Proposed Rulemaking and Supplemental Tentative Decision, *supra* Ch. 15, note 41, at ¶¶ 44–67.

[46] *Id.* at ¶¶ 125–128.

[47] Meredith and Robinson, *supra* Ch. 15, note 38, at p. 122, n. 254.

[48] *E.g.,* "Columbia Wins Contract for DoD Transpacific System," *Communications Daily*, Sept. 14, 1992 (noting that recent receipt of landing rights by Columbia Communications in Korea and Japan was a first for any separate system).

[49] "Surprises Expected in Today's Big LEO NPRM Language," *Mobile Satellite News*, Jan. 19, 1994.

[50] *E.g.,* "Orion Atlantic," *Satellite News*, Aug. 1, 1994.

[51] "No carrier shall undertake the construction of a new line or of an extension of any line, or shall acquire or operate any line, or shall engage in transmission over or by means of such additional or extended line, unless and until there shall first have been obtained from the Commission a certificate that the present or future public convenience and necessity require or will require the construction , or operation, or construction and operation, of such additional or extended line... As used in this section the term 'line' means any channel of communications established by the use of appropriate equipment ..." 47 U.S.C.A. § 214(a) (West 1995).

[52] Charles H. Kennedy and M. Veronica Pastor, *An Introduction to International Telecommunications Law*, Ch. 6, Norwood, MA: Artech House, 1996. The FCC's rules on foreign ownership under Section 214 are discussed in the Report and Order cited at Chapter 4, note 58, *supra*.

[53] Hamburg and Brotman, *supra* note 2, at § 6.02[6][c].

[54] "U.S. Satellite Companies Unwavering in Opposition to Tongasat," *Satellite Week*, Nov. 15, 1993; "Rimsat Responds to Attacks on its Use of Tonga Orbit Slots," *Satellite News*, Oct. 23, 1993; "Columbia Asks FCC to Deny U.S. Markets to Users of Tonga's Orbital Slots," *Communications Daily*, Aug. 24, 1993.

[55] Memorandum Opinion and Order, In the Matter of IDB Worldcom Services, Inc., 10 F.C.C. Rcd 7278 (1995); "FCC Approves 15 Applications for U.S. Firms to Use Russian Birds," *Satellite News*, July 10, 1995.

[56] Public Notice, Report IN 95-25, Sept. 20, 1995.

[57] Emma Tucker, "EU satellite liberalization," *Fin'l Times*, Oct. 17, 1994.

[58] Towards Europe-wide systems and services—Green Paper on a common approach in the field of satellite communications in the European Community, COM (90) 490 Final, Nov. 1990, [hereinafter "Satellite Green Paper"].

[59] *Id.* at Section VI.

[60] *Id., Short Presentation; EC Telecommunications* Law ¶¶ 2.42–2.43, 5.8-5.12, N. Higham, L. Gordon, S. White (eds.), Wiley Chancery Law, 1993 & 1994 supp.; Fernando Pombo, "European Community Telecommunications Law and Investment Perspectives," 18 *Fordham Int'l L.J.*, pp. 555, 575–76 (1994).

[61] Satellite Green Paper, *supra* note 58, at sec. VI.

[62] Council Resolution of 19 December 1991, O.J. C 8/1, 1992 (on the Development of the Common Market for Satellite Communications Services and Equipment).

[63] *EC Telecommunications Law, supra* note 60, at ¶ 5.14; Pombo, *supra* note 60, at 576–77.

[64] EC Telecommunications Law, *supra* note 60, at ¶¶ 5.17–5.22; Pombo, *supra* note 60, at 577.

[65] Pombo, *supra* note 60, at 578; Nigel Tutt, "EU predicts directive will cause satellite boom," *Electronics*, Oct. 24, 1994; "EC Ends Government Monopolies on Satellite Communications Equipment, Services," *Satellite Week*, Oct. 17, 1994; Emma Tucker, "EU satellite liberalisation, *Fin'l Times*, Oct. 17, 1994; Satellite and International," *Communications Daily*, Oct. 17, 1994; Denise Claveloux, "EC Liberalises Satellite Communications," *Electronics*, Dec. 13, 1993.

[66] *EC Telecommunications Law, supra* note 60, at ¶¶ 5.23–5.39; Pombo, *supra* note 60, at 578–79; "Euro Parliament Accepts Recognition Directive," *SatNews M2 Communications*, May 3, 1994; "Telecommunications: Commission Proposal to Harmonise Satellite Licenses," *Tech Europe*, Europe Information Service No. 90, Feb. 4, 1994; "Law Considered to Establish Europe Satellite Market," *Global Telecom Report*, March 22, 1993.

[67] *EC Telecommunications Law, supra* note 60, at ¶¶ 5.40–5.46; Pombo, *supra* note 60, at 579–80; "In the News: Key Objective of EC Satellite Policy is Region's Competitiveness," *Wall St. J. Europe*, July 18, 1994; "New Resolutions on European Telecommunications Adopted," *NTIS Alert Foreign Technology*, U.S. Dep't of Commerce, Feb. 1, 1994; Richard L. Hudson, "EC Commission Gives Wake-Up Call for Satellite Phone," *Wall St. J.*, April 16, 1993.

[68] "Today's News: European Big LEO Proposal Headed to Parliament," *Communications Daily*, Nov. 13, 1995; "Satellite and International," *Communications Daily*, Nov. 9, 1995; Julie Wolf, James Pressley, and Brian Coleman, "A Special Background Report on European Union Business and Politics," *Wall St. J. Europe*, Nov. 9, 1995; "Brussels move on satellites," *Fin'l Times*, Nov. 9, 1995.

CHAPTER 18

▼▼▼

CURRENT AND PLANNED SYSTEMS

18.1 INTRODUCTION

The purpose of this final chapter devoted to satellite telecommunications is to describe some of the systems that are currently operational or that are planned to be put into orbit by the end of the century. The ones discussed here do not constitute an exhaustive list of all systems that do or will provide two-way point-to-point telecommunications services. But there is an attempt to cover the ones that provide the largest volume of services, as well as to furnish a sample of the different forms of ownership (government, private, nonprofit, and some combination of the foregoing), the various scales of service (global, regional, and national), and the diverse technological approaches available.

18.2 INTELSAT

Approximately two-thirds of the world's international telecommunications traffic is carried by the International Telecommunications Satellite Organization, or INTELSAT [1]. INTELSAT is a multinational organization of countries and their telecommunications providers, headquartered in Washington, D.C. It was created as a global satellite communications monopoly in 1963. Membership is open to any nation that is a member of the ITU, but nonmembers may also use the space segment. As of the end of 1994, according to its annual report, it had 136 members.

INTELSAT's governing documents are the INTELSAT Agreement [2] and the Operating Agreement [3]. The INTELSAT Agreement is a treaty that sets forth the goals, scope of activities, financial principles, and governance structure of INTELSAT. It also requires each country (Party) that joins INTELSAT to sign or designate a Signatory to sign the Operating Agreement. The Operating Agreement details the operating principles of the organization. The U.S. Signatory to the Operating Agreement is COMSAT, a publicly owned corporation that was created by the Communications Satellite Act of 1962 [4]. For other countries, the Signatory is the primary or sole telecommunications operator for the country (which might also be the telecommunications regulatory authority for that country).

INTELSAT has four governing bodies: the Assembly of Parties, the Executive, the Meeting of Signatories, and the Board of Governors. The Assembly of Parties is composed of representatives of the countries that have signed the INTELSAT Agreement, and each country has one vote. The Assembly of Parties sets the long-term policy of the organization and has the power to amend the INTELSAT Agreement. It determines the structure of the Executive and appoints the Director General, who has always been from the United States. The Director General is the chief executive officer and reports to the Board of Governors. The Executive is responsible for the day-to-day affairs of the organization. The Meeting of Signatories is composed of representatives of the Signatories and its most important function is determining the level of investment share that will entitle a Signatory to have a Governor on the next year's Board of Governors. Thus, while the Assembly of Parties is composed of representatives of the countries that are Parties to the INTELSAT Agreement, the Meeting of Signatories is composed of representatives of the countries' telecommunications operators. The interests of the countries that are Parties may not always be the same as those of the telecommunications entities that are Signatories.

The Board of Governors is responsible for planning, designing, constructing and operating the space segment and for any other activities that INTELSAT may carry out under the INTELSAT Agreement or the Operating Agreement. Under the INTELSAT Agreement, the Board of Governors is to be kept as close to 20 as possible. Those Signatories entitled to a Governor are determined on the basis of investment share. Investment share, in turn, is set by each Signatory's percentage of the total satellite usage during the prior 180 days, although each Signatory is entitled to a minimum investment share of 0.05%. The countries with the largest investment share are entitled to a Governor. Countries may also, in some cases, aggregate their investment shares in order to obtain a joint representative on the Board of Governors. A group of five countries from any one of the ITU's regions may also be entitled to have a Governor named for them, subject to certain limits. Voting by the Board of Governors is weighted in relation to the investment share of the Signatory represented by the Governor.

Funding for INTELSAT is also determined on the basis of each Signatory's investment share. Each Signatory must make a capital contribution of at least 0.05% of the value of the organization upon joining. Thereafter it must make regular contributions based on its investment share as needed to meet INTELSAT's capital requirements for building and maintaining the space segment and INTELSAT's other property. Each Signatory and each other user is also billed for its use of the space segment.

A portion of INTELSAT's profits is also distributed quarterly. Each Signatory's share of profits is a factor of its investment share.

INTELSAT is the only satellite system with global universal service and nondiscrimination obligations. The "prime objective" of the organization is "the provision, on a commercial basis, of the space segment required for international public telecommunications services of high quality and reliability on a nondiscriminatory basis to all areas of the world" [5]. Certain domestic transmissions are, however, given equal priority: when two areas of a country are separated by another country or a natural barrier (e.g., a large body of water); and when geographically remote areas cannot viably be served by land-based systems due to insurmountable barriers (e.g., mountains). The space segment established to meet the prime objective is also to be made available for other domestic public telecommunications services on a nondiscriminatory basis to the extent that ability to meet the prime objective is not impaired. The INTELSAT space segment may also be used for specialized communications services such as broadcasting and space research. Finally, INTELSAT may provide certain satellites or associated facilities separate from the INTELSAT space segment provided that its operation of that segment is not unfavorably affected [6].

The universal geographic coverage goals of INTELSAT are not the only way that developing countries are helped. The pricing system is intended to provide a financial subsidy for developing countries. In setting the price of its services, INTELSAT may take into account only average overall costs. The INTELSAT Agreement provides that "rates of space segment utilization charge for each type of utilization shall be the same for all applicants for space segment capacity for that type of utilization" [7]. This means that higher rates cannot be charged for less used routes. Conversely, lower rates cannot be charged for the high-volume routes (e.g., New York to London) typically used by the industrialized countries [8].

What keeps the developed countries from simply building their own satellite systems and moving their traffic to those other systems in order to receive lower rates? In part, it is the fact that the creation of other international satellite systems is restricted by the INTELSAT Agreement. Any Party, Signatory, or person within the jurisdiction of a Party who wishes to establish, acquire, or utilize an international satellite system separate from INTELSAT (a *separate system*) must first consult with the Assembly of Parties through the Board of Governors to ensure that the new system will be technically compatible with INTELSAT's satellite network and to avoid "significant economic harm" to INTELSAT's global system [9]. (This is referred to as the *Article XIV(d) process*.) Although INTELSAT was initially resistant to allowing U.S.-licensed separate systems to operate, it has reformed this consultation process [10].

INTELSAT has jurisdiction only over the space segment. It does not construct, finance, or maintain Earth stations, although it must approve any Earth station that applies to use the INTELSAT space segment.

The first INTELSAT satellite, Early Bird, was launched in 1965. As of the end of 1994, it had in service 24 satellites, with 9 additional birds being built for launch over the following two years. It also leases capacity on satellites owned by other entities. Approximately half of its 1994 operating revenues came from international public-switched services. Another 35% were derived from transponder lease services. The remainder came from broadcast, private networks, and cable restoration. In 1996, INTELSAT plans to introduce a technology that it calls *demand assignment multiple access service* (DAMA), which it expects will make public-switched network services more affordable for developing countries served by thin (low traffic) routes [11].

In the U.S., carriers can purchase INTELSAT service only from COMSAT, the U.S. Signatory, which charges each such carrier an access charge or markup to compensate it for its role as an intermediary. This is a result of the monopoly status accorded COMSAT by the federal statute under which it was created [12].

However, some other countries have allowed non-Signatories to access the space segment. Such countries have included the United Kingdom, Chile, Argentina, and Germany. Chile and the United Kingdom have privatized the telecommunications entity that was the Signatory to the Operating Agreement, and have allowed competitors of those telecommunications entities to directly access the space segment. This means that these users do not need to work through a competing service provider. In Argentina, the Signatory wished only to perform policy and administrative functions; the provision of telecommunications service is no longer its function. Germany has also recently announced that it would allow direct access to INTELSAT facilities without investment in INTELSAT [13]. Originally, the Signatory was required by INTELSAT to be responsible for any charges incurred by or damage caused by other users, but this requirement can be waived in situations such as those described above, where it does not seem equitable for the Signatory to be liable [14].

Proposals have been made to allow multiple Signatories for a single country. The Assembly of Parties has announced its support for allowing a Party to decide whether it will allow multiple Signatories for that country. However, the country will be allowed only one vote at the Meeting of Signatories and there can be only no more than one Governor per country.

Reform of INTELSAT has extended as far as investigating the possibility of privatization. Proposals under consideration include: (1) abolishing INTELSAT's privileges and immunities but leaving it as an intergovernmental organization; (2) keeping the organization intact but eliminating government involvement; and (3) breaking the company into 3–4 global companies that would compete with each other; (4) and creation of a separate affiliate, as has been done with INMARSAT (as discussed in Section 18.3). Under a proposal by the Unites States government and

COMSAT, INTELSAT's assets, including its satellites, would be divided between the existing entity and a new, independent entity that would offer shares to the global public. However, no decision on the matter had been reached as of the date of this writing [15].

18.3 INMARSAT

The International Maritime Satellite Organization (INMARSAT), headquartered in London, England, was originally created in 1979 "to make provision for the space segment necessary for improving maritime communications, thereby assisting in improving distress and safety of life at sea communications, efficiency and management of ships, maritime public correspondence services and radiodetermination capabilities" [16]. As will be discussed in more detail below, however, INMARSAT's scope has since been expanded to include two other types of MSS, aeronautical and land mobile satellite service.

Like INTELSAT, INMARSAT's governing documents are a treaty, the Convention on the International Maritime Satellite Organization [17], and an Operating Agreement [18]. As with INTELSAT, the Convention is an agreement among nations, whereas the Operating Agreement is entered into by telecommunications entities (again, COMSAT for the United States). Indeed, the INTELSAT documents served as models for INMARSAT.

INMARSAT bears other resemblances to INTELSAT. Its governance structure comprises an Assembly of Parties, a Council, and a Directorate. The Assembly of Parties meets every two years. It expresses its views and makes recommendations to the Council on the activities, purposes, general policy, and long-term objectives of the organization. The Assembly appoints four representatives to the Council, which is composed of 22 members and is the most powerful organ of INMARSAT. (The remaining 18 members are appointed one each by the countries with the largest investment shares, although investment shares may be aggregated by a number of countries if they choose to do so.) Voting of the Council is weighted by investment share. The Council determines the organization's requirements and adopts policies, plans, programs, procedures and measures for the design, development, construction, establishment, acquisition, operation, maintenance, and utilization of the space segment. It deals with financial matters as well as the criteria for approval of equipment used in the Earth segment by users of INMARSAT services. The Directorate is the executive organ and functions under a director general who reports to the Council.

Investment share is determined on the basis of the percentage of total utilization of the INMARSAT space segment of each country. The minimum investment share is 0.05%.

The INMARSAT treaty also contains a provision on separate systems. A Party to the Convention is to notify the organization "in the event that it or any person within its jurisdiction intends to make provision for, or initiate the use of ...separate space segment facilities to meet any or all of the purposes of the INMARSAT space segment, to ensure technical compatibility and to avoid significant economic harm to the INMARSAT system" [19]. The Council is to provide its views to the Assembly on economic harm and the Assembly "shall express its views in the form of recommendations of a nonbinding nature." This is a less restrictive standard than the INTELSAT treaty, which does not use the phrase "of a nonbinding nature."

Given the fact that INTELSAT was used as a model for INMARSAT, why were the functions of INMARSAT simply not given to INTELSAT? Apparently it was felt that the purposes of INTELSAT were so broad that a specialized agency was needed to accord the necessary level of attention to the area of maritime services. Another reason is that the Soviet Union and other communist countries were not members of INTELSAT [20]. Yet, the USSR and others were major shipping nations and their participation in INMARSAT was desirable. Finally, countries that would use maritime satellites far more than they would use INTELSAT facilities (e.g., Norway) would have larger investment shares in a more limited organization than they do in INTELSAT.

As noted, INMARSAT's original function was to meet the communications needs of maritime traffic. However, the operational competence of the organization was expanded in 1985 by amendments to the Convention and Operating Agreement that authorized the provision of aeronautical satellite telecommunications. Further amendments were adopted in 1989 to confer the competence to provide land mobile satellite services [21].

The INMARSAT space segment consists of four geostationary satellites (the INMARSAT-2 series) that are owned by INMARSAT, plus leased capacity [22]. INMARSAT anticipates that its next generation of space segment technology, five satellites that will constitute the INMARSAT-3 system, will come online in 1996–97. This new system is expected to allow the size and cost of ground segment equipment to be reduced because the satellites will be more powerful.

Land Earth stations link INMARSAT's satellites with the national and international telecommunications networks. They are generally owned and operated by Signatories to the Operating Agreement, although there are some other service providers that own and operate the land Earth stations.

The following are among the services provided by INMARSAT:

- *INMARSAT-A.* The company's first mobile satellite communications offering, it provides analog service and is in the process of being replaced by its digital successor, INMARSAT-B, which will encompass the same types of services.
- *INMARSAT-B.* Terminals can be as small as the size of a suitcase, depending on the services. They are used to access basic direct-dial telephone, high-speed data transmission, multiple channel and live video, facsimile, and teleconferencing services. The equipment typically costs from $30,000–$40,000, although multiple channel versions can run up to $150,000. Service charges

generally range from $3–$7 per minute for telephony, fax, or data and $4–$7 for telex. Extra charges may apply for certain terrestrial connections.

- *INMARSAT-C*. These terminals are used to access two-way store-and-forward low-speed data or text message services. Mobile equipment costs between $3,000 and $12,000. Service charges are around $1 per kilobit of information (about 20 words) transmitted.
- *INMARSAT-D*. These terminals are pocket size and are capable of receiving and storing individually addressed or broadcast messages up to 128 characters in length.
- *INMARSAT-M*. Terminals may be as small as the size of a briefcase. They provide voice communication and low-speed fax and data transmission. The cost of the mobile equipment ranges from $13,000 to $25,000. Service charges generally are $3–$6 per minute, depending on routing and destination. Service charges may be lower for transmissions during off-peak hours.

INMARSAT estimates that the average price of all of its voice services has fallen by more than 20% in the 12 months ending in Autumn 1995.

INMARSAT services have been utilized to encourage development. Examples include a pilot scheme under which solar-powered pay phones were installed in Cameroon and INMARSAT-C was used in Nigeria for e-mail between villages.

INMARSAT has recently formed an affiliate to provide land mobile satellite service via handheld individual user units in competition with private LMSS operators. The new company was first called INMARSAT-P (for personal) but was later renamed ICO Global Communications (for the intermediate circular orbit into which its satellites will be launched). The ICO system will comprise 12 satellites (10 operational, 2 backup) orbiting at 10,000 kilometers above the Earth. The phones will use GSM cellular service where it is available, and satellite service where it is not. Service is expected to become available in about 1999 [23].

The first round of funding for ICO was obtained from INMARSAT and members that chose to participate in the new venture. The target was $1 billion, but commitments were obtained and accepted for $1.4 billion. The largest such investor was COMSAT, which contributed $147 million through a combination of direct investment of $94 million and indirect investment. (COMSAT has an interest in INMARSAT, which contributed $150 million, and wholly owns COMSAT Argentina, which contributed $20 million) [24].

Additional funding was obtained from Hughes Electronics Corporation, which took a $94 million investment share in ICO. Its subsidiary, Hughes Space and Communications International, was given a contract to worth $1.4 billion to design, develop, and manufacture the ICO satellite system. Hughes will also supply and manage launch services and will receive nonexclusive rights to wholesale ICO services in the United States (COMSAT will also act as a U.S. service provider.) [25].

COMSAT has sought approval of its investment in ICO from the FCC. Its application has been opposed by private satellite companies that will compete with

ICO. Officials of the State Department and the Commerce Department have recommended that it be denied unless certain competition-based criteria are fulfilled. Opposition is premised on the grounds that such participation could be anticompetitive. A decision by the FCC has not yet been made [26]. The question has also been raised whether ICO is so different from INMARSAT that the FCC should regulate it as an international communications service provider in its own right, rather than regulating COMSAT's participation in ICO [27].

18.4 INTERSPUTNIK

Choosing not to join INTELSAT, but rather to create its own alternative organization, the Soviet Union and eight other Communist countries signed the Agreement on the Establishment of INTERSPUTNIK in 1971; the Agreement came into force in 1972. The organization consists of a governing body, the Board, and an executive body, the Directorate. It is funded through income generated through system use and by a fund to which members contribute in proportion to their use of the satellite system. Voting on the Board is equal for each member, rather than being proportionate to usage [28].

Despite the dissolution of the Soviet Union, INTERSPUTNIK remains in operation. With its headquarters in Moscow, INTERSPUTNIK currently has about 23 members and over 100 users. It provides telephony, data, and broadcasting services [29]. The organization continues to launch new satellites [30].

18.5 EUTELSAT

The European Telecommunications Satellite Organization (EUTELSAT), headquartered in Paris, France, was formed on an interim basis in 1977 [31]. The pattern of the definitive arrangements conforms to that pioneered by INTELSAT and INMARSAT. An intergovernmental agreement, the Convention, sets up the constitution of the organization and deals with other necessary matters which are the responsibilities of the states' Parties. There is also an Operating Agreement between telecommunications entities. (A state Party may execute the Operating Agreement instead if it is responsible for operating the telecommunications services within the country of which it is the government.) The organs of EUTELSAT are the Assembly of Parties, the Board of Signatories, and an Executive Organ headed by the Director General.

As set out in the Convention, the main purpose of EUTELSAT is the design, development, construction, establishment, operation, and maintenance of the space segment of the European telecommunications satellite system(s). Specifically, the prime objective is the provision of the space segment required for international public telecommunications services in Europe. This includes both fixed and mobile services. The space segment is available for domestic services on the same basis as international services when areas under the jurisdiction of a Party are either separated by the territory of another Party or by the high seas. On a second priority basis, domestic services not meeting the criteria just identified and other international services can be provided.

Originally, when EUTELSAT was formed on an interim basis, it had 16 members; it now has 44, the latest to join being the Slovenian Republic. Membership is expected to be expanded to 48 soon [32]. To become a member, a state wishing to accede to the Convention must apply to the Director General and provide information that includes the proposed use it will make of the EUTELSAT space segment. The Board of Signatories makes a recommendation to the Assembly of Parties, which makes a decision on the application by secret vote.

At the time EUTELSAT was formed, all of its prospective members were already Signatories to the INTELSAT agreement. Accordingly, the EUTELSAT system was required to undertake the consultation process spelled out in Article XIV(d) of the INTELSAT Treaty. The finding that EUTELSAT would not cause significant economic harm to INTELSAT was based largely on the argument that Europe was heavily served by terrestrial networks, so that EUTELSAT would take customers from landline systems rather than INTELSAT. As with other separate systems, a material modification to or expansion of the EUTELSAT system requires new consultation.

The 1990 Satellite Green Paper (discussed at Section 17.3) recommended a number of adjustments to EUTELSAT. The issues are similar to those that arise in connection with INTELSAT. First, delay in the coordination process that EUTELSAT's competitors must go through before they launch their system is in the interest of EUTELSAT. Second, those who seek access to EUTELSAT's space segment will be competitors to the Signatory, and yet they can only obtain such access through the Signatory. The United Kingdom has dealt with this issue by establishing a Signatory Affairs Office that competitors may work with rather than dealing directly with British Telecom. However, to a large degree these issues can only be dealt with by amendments to the EUTELSAT treaty [33].

The majority of EUTELSAT's revenue now comes from transmission of television programming, although a substantial portion is still derived from two-way point-to-point voice communications. Its Hot Bird series, consisting of three television satellites, with a fourth recently ordered and to begin service in 1998, covers the entire continent of Europe from Iceland to Russia and reaches North Africa, Turkey, and the Near East. Its six EUTELSAT II satellites provide public telephony, business services, satellite news reporting, exchange of European Union radio and television programs, and the Euteltracs navigation and information service. EUTELSAT recently placed orders for three new Aerospatiale EUTELSAT III satellites to replace some of its EUTELSAT II birds and a fourth satellite to be built by a Russian consortium. The Russian-built satellite will cover far-eastern Europe and Central Asia, including Siberia [34].

18.6   PALAPA

The Palapa system began as a domestic telecommunications network designed to meet the requirements of Indonesia. Because Indonesia comprises a chain of islands approximately 3,000 miles long, making a wireline system problematic, and has a high rainfall density that affected terrestrial radio links, the satellite approach was logical. A request for technical coordination was sent to INTELSAT in January 1975, and the IFRB published advance information on the network in April 1975. However, there were delays in resolving technical problems and financial issues. In the meantime, other southeast Asian nations expressed interest in sending traffic over the system, so that an Article XIV(d) consultation was required. INTELSAT found that significant economic harm would not result from the Palapa system, primarily, it appears, because accessing INTELSAT facilities would be uneconomic, so that Palapa would create new traffic rather than draining INTELSAT [35].

As part of PT Telkom, the government-owned telephone company, Palapa launched six satellites: two in the Palapa A series and four Palapa B satellites. Out of these six, only three of the Palapa B's remain in orbit. In 1991, the government sold the right to operate one of these satellites, the Palapa B-1, to a company called Pasifik Satelit Nusantara [36]. PT Telkom recently submitted applications to launch two more of its own satellites, Palapa B-5 and B-6 [37].

In 1993, the government awarded the right to own and operate the Palapa C series satellites to Satelit Palapa Indonesia (Satelindo). Satelindo is owned 45% by a subsidiary of PT Bimantara Citra, 25% by a subsidiary of Deutsche Telekom, 7.5% by state-controlled PT Indosat; and 22.5% by PT Telkom. PT Bimantara Citra is led by President Suharto's second son, Bambang Trihatmodjo. The two government companies retain "golden shares" that allow them to veto any plans considered counter to state interests. The U.S Import-Export Bank has loaned a substantial portion of the money for the launch of Palapa C satellites, as it did for the A and B satellites [38].

The Palapa B system covers southeast Asia and northern Australia and provides telephone and data services as well as broadcasting. The first two Palapa C satellites are scheduled to be launched in early 1996. The Palapa C system's footprint is planned to stretch from Iran to Vladivostok and south to New Zealand [39].

Several Asian countries that have been using the Palapa system have decided to launch satellites of their own. These include Singapore, Malaysia, and Thailand [40].

18.7   ARABSAT

The Arab Corporation for Space Communications (ARABSAT), headquartered in Riyadh, Saudi Arabia, came into being in 1977. ARABSAT is an independent organization within the Arab League and, according to its founding agreement, is intended to provide communications, information, culture, education, and any other service that can be provided by it towards the fulfillment of the Arab League Charter. It has a tripartite structure: a General Assembly, a Board of Directors, and an Executive Organ headed by a Director General [41].

Many members of ARABSAT are also members of INTELSAT, bringing into play the need for consultation under Article XIV(d) of the INTELSAT Treaty. Based on information at first provided by Saudi Arabia on behalf of ARABSAT, INTELSAT's director general concluded that almost all of the traffic expected to be carried on ARABSAT's network was then being carried on INTELSAT (or was planned to be so carried). Additional information was then provided that showed plans for increased use and expansion of terrestrial links and that ARABSAT would draw traffic from these links. The finding that ARABSAT would not cause significant economic harm to INTELSAT was thus reached, as one commentator has characterized it, "by virtue of an alternative terrestrial telecommunications system which largely existed only on paper." That commentator has also stated that he is unaware of whether the projected links were fully implemented [42].

From 1985 to 1993, ARABSAT launched three first-generation satellites, some of which it leased capacity on to non-Arab institutions. By 1993, the first two had expired, and ARABSAT replaced one of them by buying an existing bird from Telesat Canada and moving it to an ARABSAT orbital slot. This satellite then had to be replaced the following year with another Telesat Canada in-flight Anik D1 satellite for fear that the first one would run out of fuel. ARABSAT intends to launch its second generation of satellites in 1996. The satellites are used to transmit voice and data telecommunications and for broadcasting [43].

18.8 PANAMSAT

PanAmSat Corporation of Greenwich, Connecticut operates the world's first privately owned "separate system" [44] As a separate system, it was required to consult with INTELSAT in order to avoid significant economic harm to INTELSAT, under the process mentioned in Section 18.2, before it could put its satellites into operation. This was a difficult process. INTELSAT had opposed the concept of private companies (in contrast to government-founded entities such as EUTELSAT) providing competitive service [45].

PanAmSat launched its first satellite in 1988. As of this writing, PanAmSat had four satellites in the geosynchronous orbit (PAS-1, PAS-2, PAS-3, and PAS-4). Its first three operating satellites combined allow it to serve approximately 97% of the world population. It plans to launch four additional satellites in the 1996–97 period and recently made an initial public offering of part of its common stock to finance two of them (PAS-7 and PAS-8). (PAS-5 and PAS-6 are being financed in large part with the proceeds of a preferred stock offering.) Its first eight satellites operate or will operate in either or both the C and Ku bands of the radio spectrum. It recently filed an application for two new satellites in the Ka band [46].

To date, PanAmSat has been prevented from interconnecting with the U.S. public-switched telephone network, so its ability to provide two-way point-to-point voice services has been limited. (The FCC's 1992 ruling that the prohibition on separate systems interconnecting with the PSTN could be lifted as of January 1, 1997, discussed earlier in Section 17.2, was in response to a request from PanAmSat [47].) Accordingly, its business to date has primarily been in other areas.

For the first six months of 1995, PanAmSat's revenues were derived from the following markets: broadcasting 67%; business communications 30%; and long-distance telephony 3%. Its business communications services include the provision of satellite capacity to domestic and regional communications carriers (particularly in Latin America) and the provision of private networks directly to end users. Because it has not been able to carry long-distance traffic between the United States and any other country, its long-distance services have to date been provided in Latin America. However, it expects to be able to offer expanded long-distance services because it will be able to carry calls to and from the United States starting in 1997, and between other countries as they deregulate and permit long-distance competition.

18.9   AMERICAN MOBILE SATELLITE CORPORATION

In 1985, the Federal Communications Commission issued a Notice of Proposed Rulemaking to allocate spectrum for mobile satellite services and to establish licensing procedures and regulatory and technical policies for what was then a new service. The Commission invited interested parties to file applications to implement MSS systems and established a cutoff date of March 29, 1985, for filing applications. Twelve applications were filed: eleven for GEOs and one for a LEO system.

In July 1986, the FCC reallocated 27 MHz of spectrum in the L band for both uplink and downlink MSS transmissions and decided that a single license to utilize this spectrum would be granted to a consortium of qualified applicants. It ruled that each applicant could demonstrate its financial qualifications by making a five million dollar cash contribution to the consortium. Eight of the applicants made the contribution. The Commission issued a license to the consortium, called American Mobile Satellite Corporation (AMSC), in 1989, to construct, launch, and operate an MSS system via GEO technology using the upper L band.

The approach taken by the FCC was challenged in court by three applicants that did not wish to make the five million dollar contribution. The FCC in 1991 issued a decision justifying its choice of a mandatory consortium approach over comparative hearings, but allowing the three challengers to join the consortium without a minimum payment (subject to the condition that their equity interest be proportional to their actual contribution) [48].

AMSC, now headquartered in Reston, Virginia, began operations in 1992 by leasing satellite capacity. It used this capacity to provide fleet management two-way mobile data messaging and position reporting services to the maritime, trucking, and rail industries. It also provides private voice network service.

AMSC then launched its own geostationary satellite in April 1995. That satellite will serve the United States, including Alaska, Hawaii, Puerto Rico, the Virgin Islands, and 200 miles of coastal waters. It is using its new satellite to continue and improve its existing services and to provide dispatch service. The company has said it will begin to use this satellite to provide mobile telephone service (voice and data, including facsimile) starting in late 1995. It will market different configurations of its service: a maritime phone to boaters and shipping operators; an aeronautical phone for corporate and general aviation aircraft users; a fixed site phone to provide basic telephone service to rural homes with no wireline service; a vehicle-mounted phone for transportation companies, construction companies, and service organizations; and a transportable phone for corporations, public safety agencies, and individuals desiring ubiquitous telephone communications availability. Satellite-only phones will be sold first. Service is expected to be available over dual-band cellular/satellite phones by 1997 [49]. The company has said that its planned rates for voice service will be $1.49 per minute, which will include long-distance charges for calls terminating in the United States, with a $25 per month access charge [50].

In 1991 AMSC filed an application to expand its GEO license to include the spectrum allocation for Big LEO systems. The FCC, however, refused to authorize GEO satellites in these bands except on a noninterference basis to LEOs. Then, in 1994, AMSC amended its application to use the Big LEO bands, proposing a 12-satellite medium-Earth orbit system to provide global handheld telecommunications services. However, it subsequently, pursuant to FCC policies, deferred its showing of financial qualifications until January 1996. It indicated that in the meantime it would continue to seek FCC authorization to use Big LEO frequencies for expansion of its geostationary domestic system [51].

The five largest shareholders of AMSC are GM Hughes Electronics Corporation (27%); Singapore Telecommunications Ltd. (13.5%); McCaw Cellular Communications, Inc., a subsidiary of AT&T (12.4%); Mobile Telecommunications Technologies Corporation (Mtel) (7%); and General Dynamics (2.3%). The remaining 37.8% is held by public investors as a result of a December 1993 initial public offering [52].

AMSC is regulated as a nondominant common carrier. It is not subject to rate of return regulation, but must offer service on a first-come, first-served nondiscriminatory basis at just and reasonable rates pursuant to tariffs filed with the FCC, which tariffs are presumed lawful [53].

AMSC's prospectus notes that several other countries are developing mobile satellite systems and that some of them (Russia, Australia, Mexico) have launched satellites for it. This raises the possibility that compatible technical standards for mobile terminals and network access may be developed [54].

18.10 VITA

Volunteers in Technical Assistance (VITA) is a nonprofit, humanitarian aid organization that provides services to developing nations around the world. In 1988, the U.S. Federal Communications Commission authorized VITA, which is headquartered in Arlington, Virginia, to construct, launch, and operate a nonvoice, nongeostationary satellite on an experimental basis [55]. The satellite was launched in January 1990 into an orbit 800 kilometers above the Earth, circling the globe 14 times per day and passing over the North and South Poles once per orbit. The e-mail capacity of the VITA satellite has been used for such applications as coordinating relief activities in Somalia and assisting doctors fighting the Ebola virus in Zaire. Also, a number of Internet gateways are or will be located around the globe, including Norway, South Africa (University of Cape Town), Chile (University of Chile in Santiago), and Australia (University of Tasmania in Hobart). Trinity College in Dublin, Ireland, recently received a license to use VITASAT for Trinet, which networks research universities and technical institutes [56].

The first-generation satellite was estimated to have a five-year life. Accordingly, in 1990, VITA filed an application to construct a two-satellite system in order to implement its complete VITASAT network, on which the FCC ruled in July 1995. VITA was able to meet the FCC's financial qualifications for the first of its second-generation satellites by entering into a contract with CTA, Inc., under which CTA would construct, launch, and provide certain operational services in return for commercial use of 50% of the satellite's capacity. VITA was not able to make the requisite financial showing for the second satellite. Under the FCC's rules, as discussed in Section 17.2, this normally would have been fatal for a Little LEO system. However, the FCC granted a waiver of its requirement that the licensee be able to show financing for two satellites and granted authority to construct, launch, and operate the first. It granted additional time for VITA to demonstrate its commitment to a second satellite [57].

Unfortunately, the attempt to launch the satellite that had been approved by the FCC failed when the rocket was destroyed during an erratic flight in August 1995. VITA has indicated that it still intends to put a two-satellite system into operation. In the meantime, it has signed an agreement to own five transponders on a small experimental satellite to be launched in 1996 by Final Analysis, which is also an applicant in the second round of Little LEO licensing, discussed in Section 18.11 [58].

18.11 FOR-PROFIT LITTLE LEOs

In addition to VITA, two other first-round applicants for Little LEO licenses have made substantial progress on putting systems into operation. The first is Orbcomm and the second is Starsys.

Receiving its authorization in October 1994, Orbital Communications Corp., or Orbcomm, was the first Little LEO system licensed by the FCC. Its license is for a constellation of up to 36 satellites, of which it has already launched two that have started providing service in North America. Extension of service to Europe and most of Latin America is planned for 1997. The entire system may be capable of operation as early as 1997, although the ability to provide service will also turn on when regulatory approvals can be obtained. In addition to position and monitoring services, the system will provide two-way messaging capability. Orbcomm will primarily sell to wholesalers rather than directly to end users. The system's owner and operator, Orbcomm Global L.P., is owned 50% each by Orbital Sciences Corp. of Virginia and Teleglobe Mobile Partners. Teleglobe Mobile Partners is in turn owned 70% by Teleglobe Inc. of Canada and 30% by Technology Resources Industries Berhad of Malaysia [59].

Starsys was granted a permit to construct, launch, and operate its Little LEO system in November 1995. The system is planned to have 24 satellites when complete. To obtain an 80% share of the company, GE Americom (a subsidiary of General Electric Co.) will spend $46 million to deploy the first two satellites. The remaining 20% is owned by a French-controlled company ultimately backed by the French government, North American Collect Localisation Satellites (NACLS). Other partners are expected to include PTTs and service providers. The company plans to begin providing service in 1997 [60].

The FCC has commenced a second round of Little LEO licensing, and a number of applicants have proposed systems. However, it is unclear whether sufficient spectrum will be available to allow these systems to be implemented. The FCC had recommended that the ITU allocate additional spectrum to Little LEOs at WRC-95, but that did not happen [61].

In addition to the United States, a number of countries are also in the process of licensing Little LEO systems. These include Belgium (Artes), France (S-80/1), Russia (Gonets), and Mexico (Leo One Pan Americana), among others [62].

18.12 IRIDIUM, GLOBALSTAR, AND ODYSSEY

In January 1995, the United States FCC granted Big LEO licenses to three applicants: Motorola, for its Iridium system; Loral/Qualcomm Partnership, L.P., for Globalstar; and TRW, Inc., for Odyssey. Iridium will be a 66-satellite system operating in the low-Earth orbit at 420 nautical miles [63]. Globalstar will also utilize the low-Earth orbit, with 48 satellites flying at 750 nautical miles. Odyssey will comprise 12 satellites in the mid-Earth orbit (5,591 nautical miles) [64]. (The higher orbit reduces the number of satellites needed to achieve the necessary coverage.)

Iridium and Globalstar have said that they expect to commence service in 1998. Odyssey plans to begin operating in 1999 [65].

One interesting technical difference between Iridium and the other two systems is that the satellites themselves will have significant onboard processing capability and intersatellite links. Thus, with the Iridium system, two subscribers anywhere on the Earth will be able to communicate with each other independent of ground-based communications networks [66]. However, when ground stations are used to perform these functions the equipment may be easier to reach if engineering adjustments must be made [67].

Iridium plans to charge end users $3 per minute plus toll charges for voice services, the highest of the three. Globalstar says it will charge a wholesale rate of .35 to .53 cents per minute and that service providers in turn will be expected to charge the public less than $1 per minute. Odyssey plans to charge less than $1 per minute plus an access fee [68]. Iridium appears to be pursuing high-income mobile subscribers. Globalstar, however, has said that fixed user terminal installations (such as phone booths) in remote villages and other sites beyond the reach of terrestrial phone networks should account for about half the traffic handled by its system [69].

The road to obtaining financing for the Big LEO systems has not always been smooth. Iridium withdrew a bond offering when potential buyers sought a higher return and greater protection than the company was willing to give. Globalstar's initial public offering of common stock received a lukewarm reception even after the company reduced the number of shares being offered and the price. Globalstar also encountered resistance to a bond offering. Investor reluctance has been attributed to doubts about whether there is a large enough market to support this number of systems (which will compete not only among themselves but with the ICO system discussed above), as well as to such factors as regulatory and launch uncertainty, among others [70].

Nevertheless, there are many willing to invest in these Big LEO projects. For example, Globalstar has used debt, vendor financing, and prepaid franchise fees from licensed operators in various countries. (Such franchisees may also own equity interests in the entity that owns the space segment [71].) Iridium has equity commitments from a group of 17 strategic investors led by a number of telecommunications service providers and equipment manufacturers around the world and is offering stakes in the consortium to developing countries [72].

Iridium and Globalstar have encountered another hurdle that may be faced by other internationally owned satellite systems. The European Commission has launched antitrust investigations to ensure that they comply with EU competition rules. ICO has already filed for such review, but Odyssey apparently has a low enough level of European investment to escape such scrutiny [73].

18.13 ECCO

ECCO will be a two-phase Big LEO system. The initial implementation is planned to consist of 11 operational satellites and 1 spare in an equatorial plane at 2,000 kilometers, to provide continuous service to all areas between 23 degrees north and south latitude (the Tropic of Cancer and the Tropic of Capricorn, respectively). Phase one does not include service to the United States. A projected second phase would involve an additional 35 operational satellites and 7 spares that would extend coverage to all areas between 71 degrees north and south latitude. Service links will be in the L and S bands, while feeder links will be in the C and Ku bands.

ECCO International will own and operate the space segment and sell wholesale capacity to gateway operators who will interconnect with the PSTN and provide service to end users. The company expects that end users will include those seeking mobile service, fixed service to their homes, and public phone service (such as solar-powered phones). Both voice and data services are to be provided.

ECCO International is currently planned as a venture between Constellation Communications, Inc., and Telebrás, S.A., the Brazilian phone company [74]. Investors in Constellation include E-Systems, Bell Atlantic Corporation, and SpaceVest. Constellation was, along with Iridium, Globalstar, and Odyssey, one of the early applicants for FCC Big LEO licenses. However, Constellation's application was deferred from consideration on financial grounds [75].

18.14 OTHER REGIONAL SYSTEMS

Other regional systems are being implemented specifically to provide service to developing or emerging countries in Asia and/or Africa. Two of them will be discussed here: Asia Cellular Satellite System (ACeS) and Afro-Asian Satellite Communications (also called Agrani). Like the American Mobile Satellite Corp. endeavor described in Section 18.9, both will use satellites in the geosynchronous orbit. Also like AMSC, they will compete with the global and some regional Little LEO systems for customers [76]. From the point of view of a national government, however, one advantage of regional systems is the ability to have somewhat more regulatory control [77].

The Afro-Asian system will comprise two geostationary satellites. Estimates of the date for the first launch vary from 1996 to 1997 (although most sources indicate late 1997), with the second approximately six months thereafter. Most recent estimates have service commencing in early 1998. The first satellite will cover mid-Asia and the second will serve Africa. They will cover 54 countries, as far west as Turkey to Singapore in the east, and from Russia in the north to Sri Lanka in the south. Phones using the system will also be able to make calls over GSM cellular networks. The system is largely owned by the Essel Group, run by Indian businessman Subhash Chandra, although other investors are being brought in and Afro-Asian plans to be publicly listed on the Bombay stock exchange [78].

Scheduled to commence service shortly after the Afro-Asian system is another that will be focused solely on Asia, the Asian Cellular Satellite, or ACeS project. The system will include at least two, and possibly as many as four, GEOs that will serve an area bounded by India on the west and by Korea and southern Japan on the east, reaching south to Indonesia and the Philippines. It is expected to achieve commercial status in mid-1998. Its customers, too, will be able to use handsets that are compatible with cellular GSM systems. The partners in ACeS are Philippine Long Distance Telephone Company, PT Pasifik Satelit Nusantara of Indonesia, and Jasmine International, a Thailand-based company. Like the Satelindo enterprise discussed in Section 18.6 in connection with the Palapa system, PT Pasifik Satelit Nusantara has as one of its shareholders a company controlled by Bambang Trihatmodjo, a son of Indonesia's President Suharto. PT Pasifik Satelit Nusantara has proposed bundling the ACeS telephone service with pay-per-view television broadcast by Palapa-C via the ACeS system [79].

18.15 KA-BAND PROPOSALS

A number of proposals are pending to utilize the Ka spectrum band, either through low-Earth or geostationary orbit satellites. Those currently being considered by the United States FCC will be summarized here [80]. It seems unlikely that they can all be licensed. However, the setting of a filing deadline by the FCC compelled companies with an interest in putting a system into operation to apply in order to preserve the opportunity to obtain spectrum and orbital position. Which ones will actually come to fruition will not be known until the FCC decides what qualifications will be imposed and whether the spectrum needed for them should be auctioned.

Perhaps the most well-known concept, both because of its scope and the name recognition of its backers, is Teledesic. The project calls for 840 or more (plus spares) low-Earth orbit satellites traveling at an altitude of 700 kilometers. Global service is planned, a goal which recently received a boost when the ITU allocated 400 MHz of Ka-band spectrum for non-GSO FSS and froze another 100 MHz for the same purpose. (Teledesic is the only non-GSO Ka-band project of those discussed in this section, although applications for others are pending at the ITU.) The satellites are to be launched in 1999 and 2000, with service to begin in 2001, although questions have been raised as to whether sufficient launch capacity is available to meet this deadline. Services are to include universal telephone service, high-speed data, and interactive two-way imaging. Authority will be sought to provide service on a mobile as well as a fixed basis. Teledesic itself will not sell to end users, but to wholesalers, including telephone companies. Teledesic was created in 1993 by Microsoft founder Bill Gates and cellular operator Craig McCaw, who recently sold his business to AT&T [81].

AT&T's application puts forth a global constellation of 12 GEOs, called VoiceSpan. Assuming a 1996 license grant, four satellites would be launched in each of the three years starting with 2000, with service to commence in the year 2000 as well. Services include basic wireless telephone service, video conferencing, electronic messaging and mailboxes, computer database access, software distribution, financial transactions, and other data applications. Internet access would be included in the range of services. The speed of data transmissions is stressed: for example, VoiceSpan will send faxes at a rate of 30 pages per minute, which is five times the pace of conventional telephone systems [82].

Comm, Inc., a subsidiary of Motorola, has filed an application for a four-satellite GEO system to be known as Millennium. Proposed service offerings include video, imaging, fax, audio, and computer data, to the Western Hemisphere. Launches would begin in 1998 [83].

EchoStar Satellite Corporation proposes two geostationary satellites to serve the United States. Proposed two-way services include high-speed switched data, video, and videophone communications [84].

GE American Communications, Inc. (GE Americom) has filed an application to construct, launch, and operate a constellation of nine GEOs (GE*Star) to serve the United States, Mexico, the Caribbean, South and Central America, Europe, India, Saudi Arabia, Pakistan, Sri Lanka, Japan, Southeast Asia, the Western Pacific, Australia, and New Zealand. Proposed service offerings include high-speed data, video, and audio [85].

GM Hughes is combining its new proposal with a prior application and is now seeking authority for a 20-satellite global GEO system, Galaxy/Spaceway, in both the Ka and Ku bands. Launching could begin as early as 1998 and would continue over the next decade, with service planned to begin in the year 2000. Service proposals include high-capacity, two-way interactive services and the ability to provide direct to the home and other forms of video distribution. Hughes has said that today's 24-minute download from the Internet would take less than four seconds at a cost no higher than today's rates [86].

KaStar Satellite Communications Corp. has filed an application to construct, launch, and operate one domestic geosynchronous satellite to serve the continental United States, Alaska, and Hawaii. Services would include voice, video, and high-speed data [87].

Lockheed Martin's proposal is for a nine-satellite global GEO system called Astrolink. Service could begin within five years of FCC approval. Voice, video, and high-speed data services would be provided primarily for business, medical, education, and industrial applications. However, service would also be sold to telephone companies, cable companies, or others who want to market it to consumers under their own names [88].

Loral Aerospace Holdings, Inc., has filed an application to expand its pending CyberStar system into a global wireless interactive multimedia transmission system having three satellites, with intersatellite links, in the geosynchronous orbit. Service would start in the United States and then be extended to other regions. Applications for the high-speed data, voice, and video communications services would include medical imaging, video conferencing, interactive computer networking, and distance learning. The company has said that the system will offer data relay at 100 times the speed of standard telephone lines [89].

Morning Star Satellite Co., L.L.C., has filed an application to construct, launch, and operate four hybrid, geosynchronous, international communication satellites. They would serve the United States (including Alaska and Hawaii), Canada, Puerto Rico, parts of Latin America, Asia, and Europe [90].

NetSat 28 has filed an application to construct, launch, and operate one GEO in the domestic FSS serving the continental United States. Proposed services include ubiquitous nationwide access to broadband communications services [91].

Orion Network Systems, Inc., and two of its affiliates have filed three separate Ka-band applications. One is for a satellite that would serve the Asia-Pacific region. Another is for a satellite that is planned to cover North and South America and West Africa. The third application seeks authorization to construct, launch, and operate three geosynchronous satellites in the FSS and to amend an existing application for a Ku-band satellite to add a Ka-band payload. The system would provide digitally compressed video, digital audio, and voice and multimedia services to the continental United States, Alaska, Hawaii, Puerto Rico, the Virgin Islands, the Indian Ocean region, locations in Southern Africa, Australia, Southeast Asia, and China [92].

PanAmSat has filed an application to construct, launch, and operate an international Ka band satellite system consisting of two GEOs: PAS-10 and PAS-11. They will serve the United States, Latin America, Europe, and Western Africa. Proposed services include a range of video programming and voice and data services [93].

VisionStar, Inc., has filed an application for authority to construct, launch, and operate one domestic satellite in the FSS, in conjunction with a local multipoint distribution system. Proposed services include an integrated local, regional, and national video programming service and interactive services, distance learning, and video conferencing [94].

Notes

[1] The discussion of INTELSAT contained in this section is drawn primarily from Charles H. Kennedy & M. Veronica Pastor, *An Introduction to International Telecommunications Law*, Ch. 4, Norwood, MA: Artech House, 1996.

[2] Agreement Relating to the International Telecommunications Satellite Organization (INTELSAT), 23 U.S.T. 3813, T.I.A.S. 7532, 1220 U.N.T.S. 21 (1971).

[3] Operating Agreement Relating to the International Telecommunications Satellite Organization (INTELSAT), 23 U.S.T. 4091, T.I.A.S. 7532 (1971).

[4] 47 U.S.C. §§ 701–757. For a detailed history of COMSAT, see Francis Lyall, *Law and Space Telecommunications*, Ch. 2, Brookfield, VT: Dartmouth Publishing, 1989.

[5] INTELSAT Agreement, *supra* note 2, Article III(a).

[6] *Id.,* Articles I(l), III(b)-(e).

[7] *Id.,* Article V. For a description of other INTELSAT activities aimed at assisting developing countries, see Alexandra M. Field, "Symposium: U.S. Trade Policy in Transition: Globalization in a New Age: Note," 25 *L. & Pol'y Int'l Bus.,* 1335, text accompanying notes 80–82 (1994).

[8] For an argument that high-volume routes do not subsidize low-volume routes, see Chris Rourk, "Analysis of the Technical and Economic Issues Raised in the Consideration of International Telecommunications Satellite Systems Separate from INTELSAT," 46 *Fed. Com. L.J.,* 329 (1994).

[9] INTELSAT Agreement, *supra* note 2, Article XIV(d). The INMARSAT and EUTELSAT treaties, discussed below, also contain consultation provisions. S. White, *supra* Ch. 16, note 1, at 116.

[10] Hamburg and Brotman, *supra* Ch. 17, note 2, at § 6.02[8].

[11] INTELSAT Annual Report, 1994; INTELSAT PSN Services: Satellite Backbone of The Global Information Infrastructure.

[12] Meredith and Robinson, *supra* Ch. 15, note 38, at pp. 94–95.

[13] "Today's News, U.S. Satellite Makers Support Privatization of INTELSAT, INMARSAT," *Communications Daily*, March 3, 1995.

[14] Field, *supra* note 7, text accompanying notes 127–132, 164–173.

[15] Gautam Naik, "Split and IPO Are Proposed for Intelsat," *Wall St. J.*, Feb. 16, 1996, p. B2; "U.S. Seeks to Have Position on INTELSAT Privatization by August Assembly Meeting," *Satellite Week*, May 1, 1995; "Government Considering Idea; Crockett: Creation of INTELSAT Affiliate May Be Politically Achievable," *Communications Daily*, April 13, 1995, p. 5; "A Talk With INTELSAT's Enzo Vitale," *Satellite News*, June 27, 1994.

[16] Convention on the International Maritime Satellite Organization (INMARSAT), Article 3. The discussion of the background and organization of INMARSAT contained in this section is taken principally from Lyall, *supra* note 4, at Ch. 5.

[17] 31 U.S.T. 1, T.I.A.S. 9605 (1976).

[18] 31 U.S.T. 135, T.I.A.S. 9605 (1976).

[19] INMARSAT Convention, *supra* note 17, at Article 8.

[20] The Soviet Union never joined INTELSAT, although Russia did become a member in 1991. Henry R. Hertzfeld, "Who's who in outer space," *USA TODAY* (Magazine), July, 1994, p. 80. Its Signatory is Morsviazsputnik. "O'Gara Satellite Networks," *Aerospace Daily*, May 18, 1995, p. 272.

[21] Wolf von Noordon and Phillip Dann, "Land Mobile Satellite Communications: A Further Development in International Space Law Part I," 17 *J. Space L.*, 1 (1989) and "Part II," 17 *J. Space L.*, 103 (1989).

[22] This discussion of the space and Earth segments of the INMARSAT system is taken primarily from the following documents provided by INMARSAT: INMARSAT Communications Unlimited, 1995; INMARSAT - B: Complete Communications in a Mobile Package, 1995; INMARSAT - C: Mobile Communications for the Information Age, 1995; INMARSAT - M: The Smallest, Lightest, Lowest Cost Global Mobile Satellite Telephone, 1995; Transat: INMARSAT - A, B, C, and M Buyers' Guide, 1995/96; INMARSAT Maritime Services, Autumn 1995; and INMARSAT Annual Review & Financial Statements. 1994.

[23] "INMARSAT-P Becomes ICO Global Communications," *Exchange*, Oct. 13, 1995; Steve Gold, "INMARSAT-P Satellite Phone Set For 1999," *Broadcast*, Oct. 12, 1995.

[24] "INMARSAT-P: A $1.5 Billion Investment Pitch," *Latin American Telecom Rep't*, Feb. 15, 1995; "Satellite Service: About Half of Signatories Join in INMARSAT-P Affiliate," *Telecommunications Rep'ts Int'l*, Feb. 2, 1995; "Raising More Than Planned," *Space Business News*, Jan. 31, 1995; "COMSAT Invests $147 Million in INMARSAT-P Affiliate," *Mobile Satellite Rep'ts*, Jan. 30, 1995; "INMARSAT-P Oversubscribes its Initial Financing by $400 Million," *Mobile Satellite News*, Jan. 26, 1995; "Telephone Co's/Satellite Phone: List of Investors," *Dow Jones News Service*, Jan. 23, 1995.

[25] "ICO Communications Signs Hughes As Strategic Partner," *Exchange*, Oct. 13, 1995; "Hughes Named 2nd U.S. INMARSAT-P Affiliate," *Mobile Satellite Rep'ts*, Oct. 9, 1995.

[26] "Administration Says FCC Should Deny COMSAT Participation in INMARSAT-P," *Mobile Satellite Rep'ts*, Oct. 9, 1995; "U.S. Government Halts Approving ICO Due to Fairness Concerns," *Satellite News*, Oct. 9, 1995; "NTIA, State Department Unsatisfied with INMARSAT-P Separation," *Mobile Satellite News*, Oct. 5, 1995; "U.S. Satellite Companies Protest COMSAT Participation in INMARSAT-P," *Mobile Satellite Rep'ts*, Sept. 11, 1995; "Satellite Dispute: TRW Petitions FCC to Deny COMSAT Application to Participate in INMARSAT Affiliate," *Edge*, July 3, 1995; "INMARSAT-P Competitors Ask FCC to Block COMSAT From Participating in Project," *Mobile*

Satellite Rep'ts, July 3, 1995; "COMSAT Defends Itself Against Industry Claims of Violations," *Satellite News,* May 29, 1995.

[27] "Long-Awaited INMARSAT-P Application Filed with FCC," *Satellite News,* May 8, 1995.

[28] Lyall, *supra* note 4, at 296–303.

[29] Hamburg & Brotman, *supra* Ch. 17, note 2, at § 6.02[5][c].

[30] "Satellite and International," *Communications Daily,* Feb. 16, 1995; "Special Report: Transponder Shortage in the United States," *Satellite News,* Feb. 21, 1994.

[31] The discussion of the background of EUTELSAT is taken in large part from Lyall, *supra* note 4, at 174–75, 264–95, which describes governance of the organization in much greater detail.

[32] "Eutelsat," *Satellite News,* June 12, 1995; "Slovenia Joins Eutelsat," *Tech Europe,* June 7, 1995.

[33] S. White, *supra* Ch. 16, note 1, at 117–132.

[34] "Space: EUTELSAT Moves Among the Major Global Networks," *Tech Europe,* Sept. 7, 1995; "EUTELSAT Orders Five New Satellites," *Tech Europe,* Sept. 7, 1995; "European Telecom AA+ Long-Term Debt Affirmed by S&P," *Dow Jones International News,* Aug. 29, 1995; "Russia Wins International Tender for Communications Satellite," *CIS Economics & Foreign Trade,* Aug. 10, 1995; "This Week's News: EUTELSAT in Big Expansion Will Spend More Than $1 Billion on 5 New Satellites," *Satellite Week,* July 17, 1995; "Aerospatiale To Build New EUTELSAT Satellites," *Dow Jones News Service,* July 10, 1995; "EUTELSAT Keeps the Same Director General for Three More Years," *Tech Europe,* May 5, 1995; "Space: EUTELSAT Seeks Tenders For 3rd Generation Satellites," *Tech Europe,* Oct. 6, 1994; "Satellite and International," *Communications Daily,* Sept. 20, 1994; Kate Maddox, "Eutelsat eyes U.S. networks," *Electronic Media,* July 12, 1995.

[35] Lyall, *supra* note 4, at 172–73.

[36] John McBeth, "Rocket Booster: Indonesian satellite concern answers critics," *Far Eastern Economic Review,* Oct. 7, 1993; "Jakarta's Methods Cloud Satellite Plan: Suharto Son's Role in Palapa Operator Casts Shadow on Privatization," *Asian Wall St. J.,* April 15, 1993; "Indonesia sells communications satellite," *Privatisation Int'l,* April 1, 1993.

[37] "PT Telkom May Be Allowed to Launch its Own Satellite," *Computergram Int'l,* Sept. 21, 1995.

[38] "U.S. Bank Finances Launch of Satellite," *Asian Wall St. J.,* Aug. 9, 1995; Richard Borsuk, "Cable & Wireless Is Disconnected From Satelindo," *Asian Wall St. J.,* June 7, 1995; Richard Borsuk, "Quiet U.S. Loan Helps Satelindo Buy Two Satellites From Hughes," *Asian Wall St. J.,* Jan 12, 1995; "Jakarta's Methods Cloud Satellite Plan: Suharto Son's Role in Palapa Operator Casts Shadow on Privatization," *Asian Wall St. J.,* April 15, 1993. The Ex-Im Bank lends money in cases where the foreign country is importing goods from the United States.

[39] "PT Satelit Palapa Indonesia to Launch Satellites Next Year," *Computergram Int'l,* Sept. 25, 1995; "Malaysia Qualcomm Systems," *Network Week,* March 31, 1995; "Indonesia - Palapa C1 Grounded," *Telenews Asia,* March 23, 1995; Manuela Saragosa, International Company News, "Foreign investors set to buy into Satelindo," *Fin'l Times,* Feb. 17, 1995; "Distance Learning Via Satellite to Grow Rapidly in Pacific Rim," *Satellite News,* May 31, 1993; Untitled Article, *Communications Daily,* Jan 8, 1993.

[40] "Singapore Joins in Satellite Space Race," *Satnews M2 Communications,* June 13, 1994; "Malaysian firm buys country's 1st satellite," *Asian Economic News,* May 23, 1994; "International Phone Update 03/18," *Newsbytes News Network,* March 18, 1992.

[41] Lyall, *supra* note 4, at 303–08.

[42] *Id.* at 175–77.

[43] "Satellite Spotlight—VSAT Technology Makes the Trip to Mecca," *Satellite News,* Oct. 17, 1994; "ARABSAT Buys Used Canadian Craft," *Satnews,* Sept. 19., 1994; "1994 Farnbourough Report, ESA Sets Reentry Test," Sept. 12, 1994; "International," *Satellite Week,* Aug. 22, 1994; "Arab Satellite TV Venture Under Way," *Screen Digest,* Dec. 1, 1993; "Telesat Canada Selling 2 Old Anik Birds to ARABSAT and Argentina," *Communications Daily,* March 15, 1993; "ARABSAT

Made $12 Million Profit," *Satellite News*, March 15, 1993; "Arab Affairs - March 2 - Used Canadian Satellite Bought," *APS Diplomat Recorder Arab Press Service Organisat*, March 6, 1993.

[44] This section is based largely on information contained in the PanAmSat Prospectus dated September 21, 1995 and on file with the United States Securities and Exchange Commission.

[45] Lyall, *supra* note 4, at 177.

[46] "PanAmSat's PAS-3 Satellite Commences Service," Business Wire, Feb. 20, 1996; "News Briefs," Exchange Jan. 25, 1996; "New Satellites: PanAmSat Targets Additional Satellites Serving Americas; Company Pursues Total Coverage of North & South America, Ka-Band Satellites for Future Telecommunications Services," *Edge*, Oct. 16, 1995.

[47] Lawrence D. Adashek, "Public and Private Antitrust Developments in the Communications Industry," 1 *CommLaw Conspectus*, 140, 147 (1993). That article describes how PanAmSat has been involved in litigation with Comsat and has sought removal of regulatory restrictions on its business.

[48] Final Decision on Remand, In the Matter of Amendment of Parts 2, 22 and 25 of the Commission's Rules to Allocate Spectrum for and to Establish Other Rules and Policies Pertaining to the Mobile Satellite Service for the Provision of Various Common Carrier Services, 7 F.C.C. Rcd 266 (1991); Tentative Decision, In the Matter of Amendment of Parts 2, 22 and 25 of the Commission's Rules to Allocate Spectrum for and to Establish Other Rules and Policies Pertaining to the Mobile Satellite Service for the Provision of Various Common Carrier Services, 6 F.C.C. Rcd 4900, 69 R.R.2d 828 (1991). The twelfth applicant withdrew its application in 1987.

[49] Jeff Cole, "Star Wars: In New Space Race, Companies Are Seeking Dollars From Heaven," *Wall St. J.*, Oct. 10, 1995, p. A1, col. 6; "American Mobile Satellite Corp. Introduces New Satellite-Based Dispatch Radio Service," *Worldwide Telecom*, Sept. 1, 1995; "American Mobile Satellite Ready for MMS Service," *Worldwide Telecom*, Sept. 1, 1995; John J. Edwards, III, "American Mobile Satellite Sees Revs $378M/Yr by '98," *Dow Jones Int'l News*, June 5, 1995; "American Mobile Satellite Corporation Satellite Launched Successfully," *Tele-Service News*, May 1, 1995; "Sprint Cellular Signs Agreement with American Mobile Satellite to Distribute Mobile Satellite Service," *Tele-Service News Worldwide Videotex*, Aug. 1, 1994.

[50] American Mobile Satellite Corporation 1994 Annual Report at 7.

[51] Big LEO Order, *supra* Ch. 16, note 13, at 5942–43, 5953; "AMSC Files for $3.1-Billion Big LEO System; Defers Financial Showing," *Communications Daily*, Nov. 17, 1994.

[52] American Mobile Satellite Corporation 1994 Annual Report at 4, 20.

[53] Prospectus, American Mobile Satellite Corp. at 25, Dec. 13, 1995.

[54] *Id.* at 41.

[55] Order and Authorization, In the Matter of the Application of Volunteers in Technical Assistance For Authority to Construct, Launch and Operate a Non-Voice, Non-Geostationary Mobile-Satellite System, 1995 LEXIS 4974, at ¶¶ 1, 3 (1995).

[56] "Satellite Spotlight: VITASAT A Vital Link in Ebola Crisis," *Satellite News*, July 24, 1995; "Volunteers in Technical Assistance," *Mobile Satellite News*, Feb. 23, 1995; VITASAT Global Electronic-Mail Gateway Network (available from VITA on the Internet).

[57] Order and Authorization, *supra* note 55, at ¶¶ 3–4, 9–16.

[58] "Satellite and International," *Communications Daily*, Nov. 20, 1995; "Mobile Satellite Diary," *Mobile Satellite Rep'ts*, Nov. 6, 1995; "VITA Seeks New Partners for Satellite Venture," *Space Business News*, Oct. 18, 1995; "Inaugural LLV-1 Launch Fails; CTA and VITA Set Back By More Than A Year," *Mobile Satellite News*, Aug. 24, 1995; "Vitasat-1/Gemstar-1 Lost; Maiden Launch of Lockheed Vehicle Ends in $17-Million Failure," *Communications Daily*, Aug. 17, 1995, p. 2.

[59] "Telcom 95," *Wireless Business and Finance*, Oct. 11, 1995; "Mobile Satellite Diary," *Mobile Satellite Rep'ts*, Sept. 25, 1995; "Teleglobe and Orbital Expand Their Plans For Satellite Venture," *Wall St. J.*, Sept. 14, 1995, p. 10, col. 2; "VITA 2nd Licensed Little LEO," *Mobile Satellite News*, July 27, 1995; "Orbcomm Plans Commercial LEO Service By End of Year," *Mobile Data Rep't*, July 17, 1995; "Orbcomm Moving Toward Commercial Service; Testing Continues," *Advanced*

Wireless Communications, July 5, 1995; "Mobile Satellite Diary," *Mobile Satellite Rep'ts*, Feb. 13, 1995.

[60] Order and Authorization, In the Matter of the Application of STARSYS Global Positioning, Inc. For Authority to Construct, Launch and Operate a Satellite System in the Non-Voice, Non-Geo-stationary Mobile-Satellite Service, DA 95-2343 (1995); "Starsys Ownership Approved by FCC, Construction Permit Granted," *Mobile Satellite Rep'ts*, Nov. 20, 1995; "GE Americom Buys Control of Little LEO Starsys, Hughes Steps Out," *Mobile Satellite Rep'ts*, Aug. 28, 1995; "Starsys Finds a White Knight in GE Americom," *Mobile Satellite News*, Aug. 24, 1995; "'Looks Like Pur-chase of License'; GE Americom Buys Control of Little LEO Starsys, Hughes Gets Out," *Commu-nications Daily*, Aug. 11, 1995, p. 3; "TRW Gets Second Patent; Starsys Passes FCC Muster," *Mobile Satellite News*, Aug. 10, 1995; "FCC Approves Starsys' Corporate Structure," *Mobile Sat-ellite News*, June 15, 1995.

[61] Report, In the Matter of Preparation for International Telecommunication Conference, FCC 95-256,¶¶ 16–26 (released June 15, 1995); "Satellite and International," *Communications Daily*, Nov. 20, 1995; "WRC-95 Decides Favorably on Several U.S. Proposals," *Mobile Satellite Rep'ts*, Nov. 20, 1995; "Final Analysis Gets Go-Ahead to Launch Second Little LEO Satellite," *Mobile Satellite News*, Sept. 7, 1995; "Space Technology: In Orbit," *Aviation Week & Space Technology*, Sept. 4, 1995; "GE Americom Buys Control of Little LEO Starsys, Hughes Steps Out," *Mobile Satellite Rep'ts*, Aug. 28, 1995; "Inaugural LLV-1 Launch Fails; CTA and VITA Set Back By More Than A Year," *Mobile Satellite News*, Aug. 24, 1995; "Orbcomm Moving Toward Commercial Service; Testing Continues," *Advanced Wireless Communications*, July 5, 1995; "Satellite's Mes-saging Ability In Question," *Space Business News*, June 15, 1995; "Little LEOs Face Domestic and International Opposition to Spectrum Plans," *Mobile Satellite Rep'ts*, June 5, 1995; "Mobile Satellite Diary," *Mobile Satellite Rep'ts*, April 10, 1995; "Mobile Satellite," *Satellite Week*, Jan. 30, 1995.

[62] "Orbcomm's Proposed Band Modifications Not Well Received By Rival Starsys," *Mobile Satellite News*, Nov. 16, 1995; "Belgium Approves Development of Two-Satellite Little LEO System," *Mo-bile Satellite News*, Sept. 7, 1995; "Belgian Government Gives Final Approval for Little LEO Sys-tem, *Satellite News*, Aug. 28, 1995; "Private, government satellite communications proliferate in Russia," *Aerospace Daily*, June 20, 1995; James R. Asker, "U.S. To Seek More Spectrum: With the 'small is beautiful' mobile satcom rapidly coming, industry's appetite is anything but small," *Aviation Week & Space Technology*, May 22, 1995; "The Universidad Nacional Autonoma de Mexico (UNAM) has reached an agreement with Leo One Panamericana," *Latin Am. Telecom Rep't*, Jan. 15, 1995; "Mexican Mobile LEO Satellite Company Says It Will Begin Operating in 1995," *Mobile Satellite Rep'ts*, Nov. 23, 1992.

[63] The name Iridium was chosen because the system was originally planned to have 77 satellites and iridium is the element with the atomic number 77. Engineers were later able to reduce the size of the constellation to 66. However, the company did not rename the project Dysprosium.

[64] Joseph C. Anselmo, "Aerospace Daily: Focus: 'Big LEO' Competitors Racing Toward Launch," *Aviation Week & Space Technology*, July 17, 1995.

[65] "New Satellites Could Overwhelm Launch Industry by '97," *Space Business News*, Oct. 18, 1995; Alan Cane, "World Trade News: Battle to offer first global mobile phone heats up," *Fin'l Times*, July 21, 1995; "MSN Special Interview: Odyssey's Bruce Gerding and Marc Leroux," *Mobile Sat-ellite News*, May 18, 1995; "McDonnell Douglas Grabs $50 Million Globalstar Launch Con-tract," *Mobile Satellite News*, April 6, 1995.

[66] Joseph C. Anselmo, "Aerospace Daily: Focus: 'Big LEO' Competitors Racing Toward Launch," *Aviation Week & Space Technology*, July 17, 1995.

[67] "In New Space Race: Medium Earth Orbit Networks," *Dow Jones News Service-Wall St. J. Sto-ries*, Oct. 10, 1995.

[68] Joseph C. Anselmo, "Aerospace Daily: Focus: 'Big LEO' Competitors Racing Toward Launch," *Aviation Week & Space Technology*, July 17, 1995.

[69] William B. Scott, "New Economics of Space: Iridium on Track for First Launch in 1996," *Aviation Week & Space Technology*, April 3, 1995; "Globalstar Executive Says Lackluster IPO Response Was Not a Set-Back, Outlines Plans for Competition," *Telecommunications Rep't*, March 20, 1995; *Communications Daily*, March 16, 1995, p. 9.

[70] Jeff Cole, "Globalstar Offering Gets Cool Reception As Investors Balk at Mobile-Phone Plans," *Wall St. J.*, Oct. 2, 1995, p. A3, col. 2; Quentin Hardy, "Iridium Pulls $300 Million Bond Offer; Analysts Cite Concerns About Projects," *Wall St. J.*, Sept. 22, 1995, p. A5, col. 1; "Project Backers Are Key; Satellite Operators Increasingly Look to Public Markets for Financing," *Communications Daily*, Aug. 24, 1995, p. 2; "Globalstar IPO Falls $82 Million Short of Company Expectations," *Mobile Satellite News*, Feb. 23, 1995.

[71] "Globalstar Executive Says Lackluster IPO Response Was Not a Set-Back, Outlines Plans for Competition," *Telecommunications Rep't*, March 20, 1995; "Globalstar IPO Falls $82 Million Short of Company Expectations," *Mobile Satellite News*, Feb. 23, 1995; "Globalstar disappoints in first LEO flotation," *FinTech Mobile Communications*, Feb. 23, 1995; "Target Was $240-$260 Million; Globalstar Completes Lower-Than-Expected $200-Million Public Offering," *Communications Daily*, Feb. 15, 1995, p. 5.

[72] "Iridium is Offering Stakes in Consortium to Foreign Entities," *Wall St. J.*, Nov. 10, 1995, p. B5, col. 5; "Lundberg: U.S. is Source of Most Regulatory Problems for Mobile Satellites," *Communications Daily*, Oct. 11, 1995; "Special Report—Iridium Files $300M Debt Offering with SEC," *Satellite News*, July 24, 1995.

[73] "Globalstar mobile satellite venture expects more questions from EU regulators," *AFX News*, Oct. 5, 1995; EC to Vet Globalstar and Iridium Under Competition Rules, *Network Week*, June 23, 1995; "In the News, EC Launches Antitrust Investigation of Globalstar and Iridium," *Mobile Satellite Rep'ts*, June 19, 1995.

[74] The foregoing information is taken from a brochure distributed at WRC-95 in Nov. 1995.

[75] "The Race to the Market Begins: FCC Licenses Three Big LEOs," *Mobile Satellite News*, Feb. 9, 1995.

[76] "See Study Finds Potential for Overcapacity in the Asian MSS Market by 2005," *Mobile Satellite News*, July 27, 1995.

[77] "Mobile Phone Systems to Approach Financial Markets," *Space Business News*, Nov. 15, 1995.

[78] "Mobile Phone Systems to Approach Financial Markets," *Space Business News*, Nov. 15, 1995; Brian Jefferies, "Satellites: The Sky's the Limit," *Far Eastern Economic Review*, Oct. 5, 1995; "Asian-Pacific Brief: Afro-Asian Satellite (India)," *Asian Wall St. J.*, Aug. 15, 1995; "Afro-Asian Satellite Communications: India's Regional Answer to the Big LEOs," *Mobile Satellite News*, Aug. 10, 1995; "Companies Spending: Stiff Competition," *Dow Jones News Service*, June 19, 1995; "Asian-Pacific Brief: TelecomAsia Corp," *Asian Wall St. J.*, May 26, 1995; Gary Samuels, "Crowded Skies: what's holding up space-based telephone service?" *Forbes*, May 22, 1995; Michael Vatikiotis and Jonathan Karp, "Upwardly Mobile," *Far Eastern Economic Review*, May 18, 1995; "Asian-Pacific Brief," *Asian Wall St. J.*, May 11, 1995, Jonathan Karp, "India/Essel/Phone Venture to Sell 20% To 30%," *Dow Jones Int'l News*, May 10, 1995; "Contracts, Financial: Afro-Asian Satellite Communications Ltd.," *Telecommunications Rep'ts Int'l*, Feb. 2, 1995; Bruce A. Smith, "HS 601 to Link Handheld Phones," *Aviation Week & Space Technology*, Jan. 30, 1995.

[79] "Technology Brief: PLDT Co.," *Asian Wall St. J.*, Oct. 13, 1995; "Satellite and International," *Communications Daily*, Oct. 10, 1995; Brian Jefferies, "Satellites: The Sky's the Limit," *Far Eastern Economic Review*, Oct. 5, 1995; "DTH-MSS?" *Mobile Satellite News*, Oct. 5, 1995; "Lockheed Martin, Ericsson Signed as Vendors for MSS Venture," *Wireless Business and Finance*, July 19, 1995; "Ericsson to Supply Dual-Mode Handheld Phones for ACeS Regional MSS System," *Mobile Satellite News*, July 13, 1995; "Lockheed-Martin Gets $650M Mobile Telephone Sys Pact," *Dow Jones Int'l News*, July 6, 1995; Martyn Williams, "Indonesian Satellite Telephone

Contracts Awarded," *Newsbyte News Network*, July 6, 1995; Michael Vatikiotis and Jonathan Karp, "Upwardly Mobile, *"Far Eastern Economic Review*, May 18, 1995.

[80]   For a list and brief description of most of these applications, see Public Notice, Satellite Policy Branch Information: Ka-Band Satellite Applications Accepted for Filing, FCC Report No. SPB-29, Nov. 1, 1995.

[81]   David J. Lynch, "Satellite Firms Aim High: Telecom giants enter crowded, high-cost race," *USA Today*, Nov. 21, 1995; "WRC-95 Decides Favorably on Several U.S. Proposals," *Mobile Satellite Rep'ts*, Nov. 20, 1995; "Satellite Diary," *Satellite Week*, Oct. 30, 1995; John Markoff, "AT&T Plan Links Internet and Satellites," *N.Y. Times*, Oct. 4, 1995; "Teledesic Envoy Details Plans for Superhighway in Space," *Exchange*, Sept. 22, 1995; "Teledesic's Better Satellite Bandwidth," *Newsbyte News Network*, Sept. 13, 1995; "FCC Covers All Bases in Proposed Ka-Band Sharing Plan," *Mobile Satellite News*, July 27, 1995; "In the News: Teledesic Faces Greater Challenges, But Offers Advantages Over Spaceway," *Mobile Satellite Rep'ts*, July 3, 1995; "6 MHz Proposed for Little LEOs; FCC Identifies Bands U.S. Will Advance at ITU for Big LEO Feeder Links," *Communications Daily*, June 16, 1995. p. 6; "'Not Serious Modification'; Teledesic Amends Application to Offer Mobile Satellite Services Outside U.S.," *Communications Daily*, Jan. 23, 1995, p. 3.

[82]   Public Notice, *supra* note 80; David J. Lynch, "Satellite Firms Aim High: Telecom giants enter crowded, high-cost race," *USA Today*, Nov. 21, 1995; "FCC Actions," *Common Carrier Week*, Oct. 9, 1995; "Internet, Video Conferencing Via Satellite the Next Hot Race," *Arizona Republic*, Oct. 6, 1995; "AT&T Planning A Global Satellite System: Includes on-line access for up to 12 million subscribers," *Fin'l Post*, Oct. 5, 1995; Bart Ziegler, Jeff Cole, and Quentin Hardy, "Satellite Plan Would Let AT&T Bypass Local Networks—If It's Ever Launched," *Wall St. J.*, Oct. 5, 1995, p. A6; Jube Shriver, Jr., "Stakes Are Sky-High as FCC Considers Satellite Service Plans," *L.A. Times*, Oct. 5, 1995, p. D-1; "AT&T Plan Ambitious; At Least 52 Ka-Band Satellites Proposed in FCC Applications," *Communications Daily*, Oct. 5, 1995, p. 3; John Markoff, "AT&T Plan Links Internet and Satellites," *N.Y. Times*, Oct. 4, 1995.

[83]   Public Notice, *supra* note 80; "FCC Officially Accepts 15 Satellite Applications," *Selected Fed'l Filings Newswires*, Nov. 1, 1995; James R. Asker, "Corporate Titans Vie for Satcom's Future," *Aviation Week & Space Technology*, Oct. 9, 1995.

[84]   Public Notice, *supra* note 80.

[85]   *Id.*

[86]   *Id.*; David J. Lynch, "Satellite Firms Aim High: Telecom giants enter crowded, high-cost race," *USA Today*, Nov. 21, 1995; "FCC Officially Accepts 15 Satellite Applications," *Select Fed'l Filings Newswires*, Nov. 1, 1995; "In New Space Race: More Alliances Seem Likely," *Dow Jones News Service-Wall St. J. Stories*, Oct. 10, 1995; James R. Asker, "Corporate Titans Vie for Satcom's Future," *Aviation Week & Space Technology*, Oct. 9, 1995; Jube Shriver, Jr., "Stakes Are Sky-High as FCC Considers Satellite Service Plans," *L.A. Times*, Oct. 5, 1995; *Communications Daily*, Oct. 3, 1995, p. 11.

[87]   Public Notice, *supra* note 80.

[88]   *Id.*; James R. Asker, "Corporate Titans Vie for Satcom's Future," *Aviation Week & Space Technology*, Oct. 9, 1995; "Internet, Video Conferencing Via Satellite the Next Hot Race," *Arizona Republic*, Oct. 6, 1995; "AT&T Planning A Global Satellite System: Includes on-line access for up to 12 million subscribers," *Fin'l Post*, Oct. 5, 1995.

[89]   Public Notice, *supra* note 80; "AT&T Plan Ambitious; At Least 52 Ka-Band Satellites Proposed in FCC Applications," *Communications Daily*, Oct. 5, 1995; "Special Report: 13 Applicants Seek FCC Nod to Use 28 GHz Band," *Satellite News*, Oct. 9, 1995; "Loral Joins Race For US Super Satellites," *Screen Digest*, June 1, 1995; "Special Report: CyberStar Venture," *Aviation Week & Space Technology*, May 22, 1995; Theresa Foley, "Mobile & Satellite: Loral heads off into broadband space race," *Communications Week Int'l*, May 22, 1995.

[90]   Public Notice, *supra* note 80.

[91]   *Id.*

[92] *Id.;* "Orion finally set for launch as it plans new fund-raising," *Telecom Markets*, Nov. 10, 1994.
[93] Public Notice, *supra* note 80.
[94] *Id.*

PART IV

▼▼▼

CONCLUSION

CHAPTER 19
▼▼▼

CONCLUSION

While it is impossible to predict precisely what share of the telecommunications markets cellular and satellite systems will take in the future, it is apparent that each will have an important role to play for some time to come. What is encouraging for developing countries is that competition among the many wireless operators and advances in technology will likely continue to drive down the price of phone service (however one defines that term) and as a result stimulate economic and social development.

The challenge for national and regional administrations will be answering the question of how these services should be licensed and regulated, a particularly difficult problem when the services provided are transnational. In this book, I have attempted to provide some guidance on how the interests of investors and governments can be balanced. As the foregoing discussion shows, a variety of approaches have been used to date in the field of wireless technology and regulation. Anyone seeking to learn from the experiences of other countries has a wealth of information from which to draw. This book can only begin to touch on it.

In writing on an entirely different subject, the English writer Max Beerbohm said [1]:

> M. Bergson, in his well-known essay on this theme says...well, he says many things, but none of these, though I have just read them, do I clearly remember, nor am I sure that in the act of reading them I understood any of them.

I can only hope that this book, despite its sometimes technical subject matter, has made more of an impression.

Notes

[1]   Max Beerbohm, "Laughter," reprinted in *The Art of the Personal Essay*, Philip Lopate (ed.), 1994, p. 239.

APPENDIX A
▼▼▼

SPECIFICATIONS OF BASES AND CONDITIONS FOR THE INTERNATIONAL PUBLIC BID FOR THE PROVISION OF MOBILE TELEPHONY SERVICES IN THE ARGENTINE REPUBLIC

*Ministerio de Economía
y Obras y Servicios Públicos*

SPECIFICATION OF BASES AND CONDITIONS

FOR THE INTERNATIONAL PUBLIC BID

FOR THE PROVISION OF

MOBILE TELEPHONY SERVICES IN

THE ARGENTINE REPUBLIC

THIS TRANSLATION IS NOT LEGALLY BINDING

Ministerio de Economía
y Obras y Servicios Públicos

SPECIFICATION OF BASES AND CONDITIONS FOR THE INTERNATIONAL
PUBLIC BID FOR THE PROVISION OF MOBILE TELEPHONY
SERVICES IN THE ARGENTINE REPUBLIC

TITLE I
OBJECT AND SCOPE OF THE LICENSE

SECTION No. 1. **Object and purpose of the bid.**

The object of the present bid is the grant of licenses to
operate the mobile telephony service in the interior of the
Argentine Republic. The mobile telephony licensees may
exercise the option to apply for other competitive services
licenses, according to their governing regulations.

Through this bid the Argentine Government seeks to obtain the
presence of operators and companies with sufficient technical
and economic capacity and with presence and desire of
permanence in the Argentine telecommunications market, to
make possible the existence of a truly competitive market, at
present with regard to this service and for the future among
all telecommunications systems. In consequence, the bidding
system has been designed so that proposals will consider the
maximun possible coverage.

The Government of the Argentine Republic, through the
mechanisms contemplated in this specification, seeks to
obtain abundant mobile communication services for the
inhabitants of the provinces, with high quality levels and
which shall be operational in the least possible time.

The Argentine Government considers that investments in
modernization of the public telephone network, made as a
consequence of the privatization of ENTel (former National
Telecommunications Company), and the technological
developments in the mobile telephony field, shall promote
improvements in the quality and price of telecommunications,
bringing them nearer international standards. Both services,
working together for the public benefit, shall be governed by
clear standards provided for in the present specification,
aimed at not altering the qualities and not distorting the
effectiveness of both services, and allowing the development
of mobile systems without stimulating taking unjustified
advantage of basic service rate distortions, which will tend
to diminish and which support an ambitious plan of
infrastructure improvements. In this sense, this bid aims at
a cellular network architecture similar to the one that would
have developed without the existence of rate distortions,

Ministerio de Economía
y Obras y Servicios Públicos

which the Government is trying to correct through a planned rebalancing and reduction of tariffs.

The Government of the Argentine Republic expects that the mobile services will serve broad rural sectors which at present have no telephone coverage giving them an unnecessary and undesirable competitive disadvantage for them.

For the purposes of the fulfillment of the objectives of the present bid, the Govemment of the Argentine Republic shall give special value to the following: a) the capacity to provide immediate service, operating the technologies presently in use in the Republic; b) the speed with which the service may be put in operation in certain urban and rural localities; c) the capacity to expand the service in order to embrace the largest population and geographic coverage; d) interconnection and coordination with the remaining communication systems, mobile services and the public telephone network; e) improved quality of service.

The scope and terms of the licenses to be granted for the operation of the service in the first band shall apply to the licensees of the second band.

In order to facilitate installation by the licensees and the extension of the service, the National Government has and will enter into agreements with entities representing the rural sector and with the provincial governments, who will make their facilities available for the provision of service to the winners of this bid. Said agreements are attached hereto as Appendix I and they shall be informed to the bidders as they continue to be entered into.

SECTION No. 2. **Definitions and Scope of the License.**

2.1 Definitions

For purposes of the services being tendered, the following definitions shall be applied:

Mobile telephony service (STM): The service based on SRMC which makes possible bidirectional simultaneous real time voice telephony communications, by means of a mobile transceiver, between two or more subscribers to said service or between such subscribers and those of the Public Telephone Networks or of other telecommunications services, either by receiving or originating communications.

Mobile cellular radio communications service (SRMC): The mobile radiocommunications service which, by means of cellular technology, permits the connection by multiple access of mobile stations among themselves and with the

Ministerio de Economía
y Obras y Servicios Públicos

national public   telephone network (RTPN). The SRMC is
part of the STM.

Cellular technology: A technology which consists of dividing
a geographic area into a number of smaller areas called
cells, to each of which a group of available radiochannels is
allocated, permitting the radiochannels used in one cell to
be reused in cells which are not adjacent, especially within
the same geographic area.

Multiple access: An operating modality by means of which a
certain number of radioelectric channels allocated to a cell
are assigned to a greater number of mobile stations,
    according to the principle of assignment as a function
of demand. Each mobile station   may   have   access
indiscriminately to any of said channels.

Mobile station or mobile transceiver (EM): A radio electric
station of the STM, intended to be used in motion or while
stationary at indeterminate points. Mobile stations include
hand-held portable units (radio electric equipment of reduced
dimensions with   their own power source), transportable
units (with their own power source) and also mobile units
installed in vehicles and other means of locomotion.

Base station (E.T.): Radio electric station of the STM, not
intended to be used in motion, used for radiocommunications
with the mobile stations, permitting them to be connected to
the appropriate C.C.M.

Mobile switching and control exchange (C.C.M.): Unit which
controls the base stations which depend on it and the mobile
units of the STM, and which furthermore provides the
switching and constitutes the interface between this service
and the RTPN.

Remote control and switching exchange (C.C.R): Unit, intended
to permit the expansion of the service, which controls base
stations which depend on it and the mobile stations of the
STM, which accomplishes mobile switching but does not
constitute an interface between this service and the RTPN. In
order to communicate with the RTPN the C.C.Rs must operate
under a C.C.M.

Cell: Area which is radioelectrically covered by a base
station or by a subsystem of the same (sectorized antenna).

Cell splitting: A technique for in subdividing the original
cells into new ones, in which the channels used in the former
are reallocated, which in order to satisfy traffic growth in
a particular geographic area.

Ministerio de Economía
y Obras y Servicios Públicos

Operating area: The geographic area awarded for the operation of the STM.

Location area (L.P.): An area within which a mobile station may move freely without the need to update the location register. It may involve one or more cells.

Location register: Updated and permanent information, available at the database(s) of a service area, containing the list of mobile stations located within the same, whether home based or roaming, as well as those home based units who are registered as roamers in another service area.

Service area: Area within which a mobile station may be reached by the subscribers of the RTPN, without them knowing the real position of the mobile station within the same.

C.C.M. area (ZCCM): Location area or group of these operated by one same C.C.M.

Mobile station home area: Location area in which a mobile station is registered as being "home".

National STM network: Combination of service areas accessible by mobile stations which are totally compatible with, and authorized for the STM, within the bounds of the national territory.

STM network: A combination of C.C.M. areas, which are part of a common numbering and routing plan.
Mobile subscriber: Registered user of the mobile station who has subscribed to STM.

Mobile subscriber roaming service: Service provided for roaming subscribers outside of their home area, when in transit through location areas in which they are not registered as home base users.

Call transfer between cells (hand-off): Automatic transfer of a call in progress from one cell to another, in order to pemmit established calls to continue, as mobile stations move from one cell to another.

Telephone service company: Company in charge of the provision of basic telephone service in the recognized operating area, as may correspond in each case.

STM operating entity: Compant responsible for operating the STM in a particular operating area.

Control channel: Channel which is used for signalling and control, including hand-off, for the transmission of digital

Ministerio de Economía
y Obras y Servicios Públicos

control information from the base station towards the mobile station and vice—versa.

Voice channel: Channel used for voice conversation and by means of which, moreover, digital messages may be sent from the base station to the mobile station and vice-versa.

CNT: Comisión Nacional de Telecomunicaciones. (National Telecommunications Commision)

LSB: Basic telephone service licensees, Telefónica de Argentina S.A y Telecom Argentina Stet-France Telecom S.A.

RNI: Reglamento de Condiciones y Normas de Interconexión. (Interconection Regulations)

RTPN: National Public Telephone Network belonging to the companies which provide basic telephone services.

2.2. Applicable technology

Considering the existing level of technical evolution in this matter, at this time the current technological solution for the provision of mobile telephony services shall be considered as that defined as "mobile cellular radio communications service (SRMC)", subject to the technical standards which govern it (AMPS and/or N-AMPS), without prejudice of the rights of those who become licensees to use, when available, other technologies and standards, compatible with the system and consistent with the scope of the license.

2.3. Frequency bands

The service which is the object of this bid shall be provided using the first frequency band allocated to the SRMC, called the "B Band", which embraces 835 to 845 Mhz for transmission from the mobile station and 880 to 890 Mhz for transmission from the base station. When justified for traffic demands, the so called "B'-Band", which embraces 846,495 to 848,985 Mhz for transmission from the mobile sation and 891,495 to 893,985 MHz for transmission from the base station, may be also used. If the CNT were to resolve to modify frequency allocations for the service in the future, the Argentine State would compensate the licensees for the cost that such a change may imply, except in the case that such allocation responds to a request made by the licensee.

2.4. Operating areas tendered

Ministerio de Economía
y Obras y Servicios Públicos

By means of the present tender it is intended to award independent licenses for the provision of mobile telephone services in Operating areas I and III of the Country, outlined in Appendix II to this Specification of Bases and Conditions.

SECTION No. 3. **Applicable Legal Standards and Provisions**

3.1. The Legal Standards and Provisions which are applicable for the purposes of this Public Bid and for the Licenses granted as a consequence of the same, shall be:

A) Act No 19.798 Ley Nacional de Telecomunicaciones (Telecomunications Act).
B) Act No 23.696.
C) The Resolution and Decree approving the present Tender Specification.
D) Decree No. 663/92 and amendment.
E) Decree No. 506/92 and amendment.
F) Decrees No 731/89 and its amendment Decree No 59/90.
G) Decrees Nos. 62/90 and its amendments 420/90, 575/90, 636/90, 677/90 and 1130/90.
H) Decree No. 1185/90, and amendments.
I) Decree No 2332/90.
J) Decrees No 2334/90 and 2347/90.
K) The other Act 19.798 regulatory Decrees.
L) As long as the cellular mobile radiocomunications service (SRMC) is the technological solution in force, the provisions of Resolutions No. 498 SC/87, 473 SC/87, 695 SC/87 (excluding the Appendix to Standard SC.02.61.,01) and its amendments, as well as resolutions Nos. Nos. 299 CNT/92, 316 CNT/92 and its concordants.
M) The CNT Resolutions allowing the use of new technologies considered compatible with the system and consistent with the scope of the license.
N) The RNI which shall include the provisions established in this specification.

The above listing does not imply any priority order, except for those cases of expressed revocation and without prejudice of the applicability of general principles relating the priority of standards.

3.2. All provisions in this specification referring to licensees, shall be applicable to the pre-awardees.

Ministerio de Economía
y Obras y Servicios Públicos

TITLE II
LICENSE CONDITIONS

SECTION No. 4. **Rights of the licensees.**

4.1. <u>Period of exclusivity</u>

For the purpose of section No. 7 of Decree No. 506/92, the Argentine National Government stipulates as one year, to be counted from the date of execution of the preaward contract, without possibility of question in this regard nor proof to the contrary, the term in which the pre-awardee of the bid shall be in acceptable condition to operate the service. As a result, the National Government hereby decides, using its attributes and in conformity with provisions in Decree No. 62/90, that the licensees of the second band (band "A") of the STM, which shall be independent companies owned by the LSBs, may not operate the service in competition with the licensees which arise as a result of this bid, until the passage of two years counted from thirty (30) days from the expiry of the term to receive objections, in the case that there are no objections, or from the date on which they are resolved, at which time the Subsecretaria de Comunicaciones (Undersecretariat of Communications) shall execute the respective pre-award contracts for the provision of mobile telephony services in each of the operating areas.
With the exceptions provided for in Decree 62/90 as amended and Decree 506/92 as amended regarding the operation areas in which the band has been divided, second band licensees shall be under the obligation of operating the service according to the same conditions as those governing the first band licencees, enjoying the same rights and under the same obligations in respect of the total population coverage, mandatory area coverage, population coverage outside the mandatory areas, coverage of linear kilometers of established roadways and speed in operation of the services according to a fixed quality. Licencees of the second band shall be subject to the same penalties and conditions governing license revocation as the first band licencees are.
Execution of the preaward contract shall imply permission for the start up and provision of the service being tendered and the right of the pre-award winner to obtain a definitive license one year after the date of said contract.

4.2. Once the licenses have been pre-awarded, their service plans shall be immediately communicated to the corresponding basic service licensees, according to provisions in Decree 62/90, Annex I, section 10.4.7, so that they may proceed to take the necessary measures to permit the correct operation of the service within the scheduled periods. If the terms which were stipulated in the pre-awardee's bid for the provision of service expire and the service cannot be

Ministerio de Economía
y Obras y Servicios Públicos

provided under the conditions specified, for reasons
dependent upon the basic service licensees, the period of
exclusivity shall be extended for the same number of days as
it may take to overcome such matters. For the purpose of
counting de number of days of the delay, the terms
established by section 8.7., Annex I, Decree 62/90, and the
reasonable term established by the CNT if that were the case,
shall be considered for the construction to be made by the
STM licencee.

4.3. Other competitive services

The award of a mobile telephony license shall not impede the
licensees from providing any other competitive service, in
accordance with the regulations which may be handed down by
the CNT for each of them. In the case of link concentrator
services (trunking), the pre-awardees of the licenses shall
be entitled to the allocation of frequencies, which shall be
applied for on a city by city basis, in the 800MHz band
(generic name), according to availability and in accordance
with the rules governing the service.
Coverages by means of these or other services shall not be
considered for the evaluation of proposals.

4.4. Usage of links for other services

In accordance with the provisions of Section 27, clause b) of
Decree 1185/90, licensees may use their own links for the
provision of competitive services, when duly authorized by
the authorities and in conformity with current normative
standards, the resale of same being prohibited, during the
exclusivity period established under Decree No 62/90, Annex
I.

4.5. Mobile public telephones

The licencees shall be entitled to install cellular or other
mobile technology public mobile transceivers

SECTION No. 5. **Obligations of the Licensees.**

5.1. The licensees are under the obligation to ensure the
continuity, regularity, equality and generality of the public
services they provide, being forbidden to make discriminatory
arrangements not providad for in this specification, or
subsidize other telecomunication services providers.

5.2. Conditions of the service

5.2.1. For the purpose of complying with the standards and
recommendations inherent in the provision of the services,

*Ministerio de Economía
y Obras y Servicios Públicos*

their quality and use of the rules of best practice. the licensees shall satisfy the provisions of this section, as well as domestic regulations and international treaties, accords and agreements in which Argentina takes part.

5.2.2. Fundamental technical plans and applicable technical standards and those established by the CNT , concerning operating compatibility, minimum service quality and network interconnection, shall be observed.

5.2.3. Signalling shall:

a) be adapted to the signalling systems presently in use by the RTPN in the service areas. Buyers of the specification shall be furnished with governing systems in cities of more than one hundred thousand (100.000) inhabitants.
b) be planned and equipped with the R2 signalling system, as it is defined in the Fundamental Signalling Plan.
c) be planned and equipped to provide the CCITT No. 7 signalling system, to be implemented in the future.

5.2.4. The STM numbering plan shall comply with recommendation Q.70 of the CCITT.
The STM licensees may enter into agreements with the basic telephone service licensees covering other numbering plans and access systems which shall be submitted to the CNT for approval.
Numbering facilities and their evolution provisions shall be established in the interconnection agreement concluded between the parties; the periods for their provision being those established by section 8.7.1, Annex I, Decree 62/90.
The plan shall be submitted to the CNT for approval. The CNT shall ex-oficio decide on the matter or, should disputes arise bettween the parties, at the request of any of them.
The numbering plan approved by the CNT shall be implemented within a reasonable period of time allowing the networks involved to be adapted, and shall be available to all operators at the same time.

5.2.5. The service shall forsee the adaptability to recommendation 687 and report 1153 of 1990 issued by CCIR concerning future public mobile telecommunication systems (FSPTMT).

5.3. Monitoring and control by the CNT

Each licensee company shall have available the instruments and equipment necessary for the CNT to exercise its faculties of monitoring, control and verification of the compliance of the service in matters of quality and quantity. Furthermore, it shall grant CNT staff duly accredited before the licensees, free access to the computer installations

Ministerio de Economía
y Obras y Servicios Públicos

necessary to effect such control. On the other hand, it shall inform the CNT about all service matters required, within the term stipulated in each case.

The CNT may inspect the licensees installations at any time, as well as the processes of installation and/or provision of the service in any of its stages and/or may require the provision of the information in this regard.

The licensees shall furnish the CNT, at its request, with digitalized maps of the service areas in order to compute coverage calculations.

5.4. Rights of the subscribers

The licensees shall provide adequate mechanisms for the receipt and attention of their subscribers complaints. Subscribers shall be entitled to have free access to an assigned number for the attention of their complaints. At the same time they shall observe the service regulations which stipulate the conditions for service provision and the rights of the customers, which shall be approved by the CNT in consultation with the licensees.

5.5. Mandatory service coverage.

The pre-awardees of licenses shall provide mobile telephony services in the service areas offered in their proposals. Within their operational areas, coverage shall be mandatory according to scheduled times, in urban areas within the political limits of the cities given in the following chronogram:

Areas I and III	start up (in months following the execution of the contract)
ROSARIO	14
CORDOBA	14
SANTA FE/PARANA	20
TUCUMAN	17
RESISTENCIA/CORRIENTES	23
MENDOZA	17
MAR DEL PLATA	20
NEUQUEN/GRAL. ROCA	23

5.6. Investments in infrastructure

Each licensee shall undertake for their account and risk, in the Area involved, the necessary investments for the construction and operation of the necessary networks for the provision of the service in compliance with the provisions of

*Ministerio de Economía
y Obras y Servicios Públicos*

the present Specification, including those in equipment,
materials, infrastructure works and installations which cover
the mandatory service areas and all others to which they bind
themselves in their proposal.

5.7. Tracking of demand

Each licensee shall undertake the necessary investments to
provide for orderly future growth of the demand in the entire
territory corresponding to the area awarded.
The obligation to track demand shall be limited by the
economy of the necessary investments, by the technological
possibilities and by the existence of an operator in the
competing band.
The service shall be provided following the demand in those
areas where there is coverage having a maximum waiting period
for service connection of four (4) months, to be counted from
the date of application for the service. In order to control
the compliance with this obligation, a biannual report
stating the number of applicants and connections, shall be
submitted to the CNT.

5.8. Service timeframes

The licensees shall complete the works and provide the
service on the dates stipulated in section 5.5. or on those
given in their proposal, if they were earlier or if they
refer to service areas which are not contemplated in said
section. This obligation is independent of the one
established in the above section.

5.9. Reports and timeframes for completion of construction

During the first three (3) years counted from the date of the
pre-award, licensees shall biannually submit, a report on the
progress of their current projects and installation plans
with respect to the one given at the time of their proposal.
In those cases where delays occur due to force majeure or
acts of God, they shall substantiate the causes which
originated them, the corrective steps taken to recover the
time lost and the new start up dates planned for the service
areas contracted. The acceptance or rejection of the
presentation made in this regard by the Licensees involved,
shall be subject to resolution by the CNT, who, if
appropriate, shall apply the procedures provided in
section 25.2.

5.10. Release of bands

The licensees shall undertake, at their own expense, all
procedures necessary, where applicable, to obtain the release
of the currently occupied bands allocated to the service and

Ministerio de Economía
y Obras y Servicios Públicos

which were awarded to them. and providing an equivalent facility at market costs and within a reasonable term —which eventually will be established by the CNT— to legally authorized users.
Illegal users shall be decommissioned by the CNT.

5.11. Emergency services

The licencee shall facilitate, free of charge, access to emergency and disaster services, specifically to police, ambulance, fire brigade and, when appropriate, civil defence numbers.

SECTION No. 6. **Independent providers**

6.1. Effective as from one year of the award of the license, for a certain area not being served by one or both bands:

A) If a third party were to offer to operate the service, the CNT shall so notify the Licensees for the operating area in a reliable manner. If they do not report within sixty (60) days of receipt of notification their willingness to install a
        network in said area and to operate the service in a reasonable lapse of time, which in no case may exceed twelve (12) months, the CNT shall award the area to the interested third party. In this latter case, the conditions governing the license awarded shall be those valid for the licensees of the operating area with respect to the area awarded.

B) If the CNT proved, based on substantiated documentation and studies, that the area justifies the implementation of a network and the consequent operation of the service, it may require the provision of said services by the licensee or licencees of the operating area. If they do not report within sixty (60) days of receipt of notification their willingness to install a network in said area and to operate the service in a reasonable lapse of time, which in no case may exceed twelve (12) months, the CNT shall proceed to issue a new call to tender, exclusively for that area.

6.2. The foregoing provisions shall not become applicable to the service areas offered, until such a time as the terms mentioned in the bid or stipulated in the plan of the second band, have expired.

SECTION No. 7. **Concerning the Networks to be installed.**

7.1. The network's architecture shall adapt itself in principle to the example given in Figure 1 of part "A" of report 1156 of 1990 CCIR and Figure 4 of report 1153 of 1990 CCIR; at the same time it shall provide for the adaptation to the scenario described in Figure 5 of report 1153 of 1990

Ministerio de Economía
y Obras y Servicios Públicos

CCIR. However, network architecture may evolve to more complex stages allowing the multiple use of carrier services of any kind, in accordance with the provisions in section 7.3., in the specification.
During the exclusivity period provided for by Decree 62/90, Annex I, the network architecture shall plan, at least, the installation of one CCM at each of the mandatory cities mentioned in section 5.5. During that period, the installation of a ET reporting to a CCM shall be prohibited within the area of another CCM.

7.2. Network changes

All changes in the mobile telephony networks to be installed which are included in the proposal submitted by the licensees, which does not negatively alter the coverage and the times of start up foreseen in the bid, or all changes in the ones already installed which do not affect the unallocated frequency spectrum, shall fall under the exclusive responsability of the licencee, and can be implemented subject only to the obligation of prior communication to the CNT, which may reject it in the event of incompatibility with the service, violation of the latter subsection or inconsistency with the scope of the license.

7.3. Links to be provided during the exclusivity period

7.3.1. Licensees shall be free to use the types of links they find most convenient, being entitled to install and use their own networks and links, connecting ET with CCM, ET with CCR, CCR with CCR, and CCR with CCM, within their oprating area. Licencees may established these links, by renting point to point links from the LSBs.

7.3.2. During the exclusivity period defined in Decree 62/90, Annex I, links connecting CCMs among themselves and with the RTPN shall be provided by the LSBs with the terms and effects provided for in section 8.7.1., Annex I, Decree 62/90, and facilities shall be furnished in quantity and capacity sufficient to meet the traffic demand under non discriminatory policies.

7.3.3. During the exclusivity period defined in Decree 62/90, Annex I, for the purposes provided for in section 7.3.2., STM licencees shall have the right to require from the LSBs, the provision of links between CCMs. Provided that price and quality are equal, which shall be established by the CNT in case of disputes, the LSBs may offer these links through the RTPN.

7.3.4. Ecxept for the provisions in section 7.3.5. of the Specification, in the provision of the links mentioned in

Ministerio de Economía
y Obras y Servicios Públicos

section 7.3.2. the regime established by Decree 62/90 Annex I. section 10.4.7. and Decree 1185/90, sections 27 and 28 shall apply, and in accordance with it, the parties shall agree the terms and conditions of provision. Should parties fail to reach an agreement, charges shall be determined by the CNT having regard to the following principles:
a) Charges shall not be discriminatory.
b) Charges shall be designed to cover service provision costs, including a reasonable return on the fixed assets employed, due to provision of the link.

7.3.5. As an exception to the general principle established in the above section, communications between mobiles and between a mobile and a fixed subscribers of different operating areas of the STM, shall enter the RTPN via the CCM which is nearest to the origin of the call. STM licensees and LSBs shall agree the charges for the use of the RTPN beteen the CCM of origin and the call destination, on the basis of the usual cost of communications through the RTPN, less a rebate. Should parties fail to reach an agreement, the rebate shall be determined by the CNT taking into account:
a) Savings obtained by the LSBs through the participation of both systems in completing the calls.
b) STM licensees billing costs.
c) STM licensees financial and bad debt costs.

SECTION No.8 **Changes in technology.**

8.1. The application of new technologies approved by the CNT which are compatible with the service and consistent with the scope of the license, shall be the licensees' right, within the limits of the present section. In no case may the usage of new technologies diminish the quality of the service.

8.2. For these purposes, the technologies and standards approved by the Federal Communications Commission (FCC) of the United States of America and the panamerican standard established under the Mercosur Treaty and under multilateral or bilateral international agreements, shall be considered fully applicable in the Argentine Republic. The CNT shall declare such applicability depending on whether they are compatible with the general standards on telecommunications in the Argentine Republic, with the scope of the license, and with the allocation of radio spectrum in force at that time.

8.3. In the event that the FCC has not approved technologies or standards which the licensees deem applicable to the services under tender, they shall prove that the same are compatible with the service and consistent with the scope of their license. Once the CNT has verified that they are compatible with the general telecommunication standards in the Argentine Republic and with the allocation of radio

Ministerio de Economía
y Obras y Servicios Públicos

spectrum in force at that time, it shall proceed to issue the pertinent authorization.

8.4. Digitalization, spread spectrum and personal communications services (PCN or PCS) technologies shall be considered specifically applicable, provided that they do not violate the exclusivity in the provision of the services defined by section 8.1., Annex I, Decree 62/90 as amended, which may be used in the frequency band allocated to the service and tendered in this Specification.

8.5. Licensees may provide services using new technologies on frequencies other than those being tendered, only if they prove, and if the CNT approves, that said technology, persuant to the panamerican standard, is effectively usable in the new proposed band.
In this case, the CNT shall allocate the new band to the service and require the licencee who applies for the allocation of another frequency to choose, within a reasonable term, between it and the one awarded by means of this tender, for the purpose that such an application shall not limit the possibilities of competition.

8.6. In order to file an application for the use of new technologies or frequencies, this procedure is to be followed:

A) The licensee shall submit to the CNT, at least three (3) months in advance, an application for an additional frequency or a change in technology and/or the amendment of the technical standards which regulate the service.

B) The licensee shall include in its presentation sufficient technical justifications, including an analysis of the compatibility with the service and consistency with the scope of the license, a study of solutions used to deal with similar problems in other world markets and of possible other technological and technical alternatives available. If it is applicable, it shall include recommendations for the amendment of current standards, attaching thereto available background information from other markets.

8.7. After evaluation of the submission referred to in the above subsection, The CNT shall, after hearing all STM licensees, authorize the use of the new technology, establishing reasonable terms for their adaptation to the system's global compatibility.

8.7.1. When the application of the new technology implies the use of a frequency band different from the one awarded in this tender, the following guidelines shall be followed:

Ministerio de Economía
y Obras y Servicios Públicos

a) the other STM licensees shall exercise the option of using the new technology within a period not in excess of three (3) years from the CNT s decision, or five (5) years from the award of the license, whatever is earliest. In no case the period to exercise the option shall be less than one hundred and eighty (180) days, term that shall be applicable after the five (5) years above mentioned.

b) During the option period, services in the new band shall not be called for bidding.

c) The CNT shall establish the guidelines to enable, in the shortest possible term, the tansition from one technology to the other and the use of the new system by the licensee.

8.8. Even if there is not an application from a licensee, the CNT shall be entitled to bid new STM services in other bands after the expiration of the terms mentioned in the above section.

SECTION No. 9. **Concerning the Subscribers.**

9.1. Resident Subscribers (home-based).

All of those residing in the service area served by the licensees of a particular operating area, having equipment meeting the requirements of the CNT and who fulfill the requirements for connection to the system, may become subscribers to the mobile telephony service of said area. The subscribers shall be fully responsible for the use they make of the service, as well as for the equipment they own.

9.2. Roaming Subscribers

Temporary connection shall be permitted, for continuous periods not in excess of six (6) consecutive months, of equipment used by subscribers to services in other operating areas of the Country and of third countries.
In order to permit such a temporary connection, the licensees shall determine that the interested party is an active subscriber of another service with whom the licensee has a current agreement covering roamers. Such verification may consist of the presentation of a copy of the last paid invoice issued by the service provider in which they are resident subscribers.

9.3. Subscriber Files of the CNT

Within the first ten (10) working days of each month, the licensees shall submit to the CNT, in the form of an affidavit, a written report giving the following data corresponding to the previous month:

*Ministerio de Economía
y Obras y Servicios Públicos*

a) subscriber additions and deletions
b) temporary disconnections and reconnections, whether at the subscriber s request, or for delays in excess of thirty days in payment.
c) the net subscriber base at close of the previous month.

9.4. Connectivity of subscriber equipment

The licensees' responsibility with regard to permanent or temporary connections of subscriber equipment, shall be limited to determining that, whether they be permanent or roamers, they comply with the conditions for use within the country as prescribed by the CNT or by current international agreements, between countries or service providers.
Except for provisions related to roaming subscribers, the licensees shall deny the provision of service if the subscriber equipment is not approved by the CNT for use within the ` Country and/or the applicant does not satisfactorily prove ownership and/or the legal presence within the Country of same and they shall accede to the connection of the equipment in the contrary case.

9.5. Sale of Subscriber Equipment.

The sale of subscriber equipment which complies with the requirements of the CNT is deregulated, is free and may be undertaken by the service provider or by any other vendor, at the subscriber's option. Licensees shall be forbidden to establish priviledge mechanisms limiting their competitive supply.

SECTION No. 10. **Concerning the Levies, Fees and Duties applicable**

Licensees shall pay the CNT the radio electric fees and duties based on the number of subscribers in service, in accordance with the provisions established in Resolution 147 Sub.C/90.
With respect to the levy of section 11 of Decree No. 1185/90, covering control, monitoring and verification, equivalent to fifty one hundredths of a percent (0.5%) of the total revenues arising from the provision of services, net of taxes and fees which are applicable, save this one, the licensees shall comply with the regulations to be handed down for this purpose.

SECTION No. 11. **Interconnections between networks and contractual relationships between Mobile Telephony licensees among themselves, and/or between them and the basic service licensees and/or the Independent Operators of Basic Telephone Services and/or with other Telecommunications Services.**

Ministerio de Economía
y Obras y Servicios Públicos

11.1 The networks of the mobile telecommunication and basic services shall be interconnected. It shall be the obligation of the Mobile Telephony Licensees to foresee the interconnectivity between the networks, by means of appropriate contractual relationships, between each other and with the basic telephone service licensees and/or the independent basic service operators and/or the other telecommunication services, in accordance with the provisions of the "Reglamento de Condiciones y Normas de Interconexion" (RNI - Regulations Governing Interconnection Conditions and Standards), approved by CNT, and which shall observe the guidelines established in this Specification in relation to this service.
Provision shall also be made to ensure compatibility of the systems for purposes of interconnection, in accordance with guidelines to be set by the CNT.

11.2. STM licensees shall channel the traffic which originates in their systems through their own means, and through those of the other providers of telecomunications services as appropiate, in accordance with section 7.3 of the Specification.
Should facilities of other service providers be used, the interconnection agreements mentioned in section 11.1. shall be concluded.

11.3. Terms and conditions of interconnection shall be agreed between the parties.
Interconnecton agreements shall determine, among other items, the operational arrangements for establishing and maintaining interconnection; the interconnection points; the interconnection capacity; the way to carry the signals between both networks; the quality of signal transmission; the arrangements to carry national and international long distance signals; telecommunication and information services required to guarantee that interconnection points shall be established and maintained; contigent obligations; dates or periods to make interconnection; the charges; and other items agreed by the parties or established by the RNI.

11.4. Should disputes arise beteen the parties, and at request of any of them, the CNT shall make a determination taking into account the following principles:
a) STM licensees shall be considered providers of competitive telephony services, and so they will be considered in the interconnection agreements.
b) Providers of telephony services or other interconnected telecommunications services, both mobile and fixed, shall recognize the services provided by the other system, according to the standards provided for in the RNI and this Specification.

❦

Ministerio de Economía
y Obras y Servicios Públicos

c) During the exclusivity period defined in Decree 62/90,
Annex I, the interconnection terms and conditions shall take
into account the rights and the consequent obligations of the
LSBs.
d) Interconnection terms and conditions shall include no
mechanisms producing inefficient use of networks and
equipments belonging to providers of telephony services or
other services.
e) STM providers shall be entitled to choose reasonable
interconnection points with the RTPN, taking in consideration
the service demands.
During the exclusivity period defined in Decree 62/90, Annex
I, interconnection links shall be provided by the LSBs
according to availability. Should access to the RTPN be
unavailable, terms and effects provided for in section
8.7.1., Annex I, Decree 62/90, shall apply.
f) Prices of interconnection provided by the LSB shall
reflect the criteria provided for in the other standards of
this Specification, in section 16 of Decree 731/90, in
section 10.4. of Annex I of Decree 62/90 and in the RNI.
g) The clauses Established shall not be discriminatory.

11.5 The CNT shall review the terms, conditions and prices of
interconnection, and shall intervene in accordance with
Decree 62/90, Annex I, section 10.4. and with Decree 1185/90,
section 27 b) and 28, being entitled to require an amendment,
should any violation to the applicable legal standards arise.

11.6. The STM licensees may enter into agreements with third
parties for the provision of service, with the objective of
enlarging geographic coverage offered, for the purpose of
providing it for account and order of the party of the first
part, under the conditions which they may agree upon. Such
agreements shall be approved by the CNT.

11.7. STM licensees competing or which should compete within
one operating area, may enter into agreements among
themselves for the joint usage of installations or equipment,
with the CNT's prior authorization, which shall safeguard
the maintenance of the conditions of free competition between
the STM service providers.

SECTION No. 12. **Concerning Prices and Tariffing**

12.1. Mobile telephony services prices shall be free and be
the sole responsibility of the licensee, within the
provisions of section 24.2..

12.2. Pricing and tariffs shall be general in nature and nondiscriminatory, and the original tariffs and subsequent
amendments shall be communicated to the CNT, simultaneously
to their publication and effective date.

Ministerio de Economía
y Obras y Servicios Públicos

The CNT may authorize lower differential general prices for small localities.

12.3. By exception, in the event that monopolistic behavior is observed, generating abusive tariffs which bear no relation to the investment effected and the cost of the service, the CNT shall be entitled to declare the existence of such monopolistic behaviour and to require the submission of new prices for approval and, when appropriate, shall proceed to fix them.
In order to support its decision the CNT must prove:

a) The existence of monopolistic behavior.
b) The non-existence of effective competition, considering all the telephony services.
c) An unreasonable relationship between the investments effected, their amortization and rate of return, with the proposed tariffs.
d) An unreasonable relationship between the service costs and the proposed tariffs.
e) An unreasonable relationship with international tariffs currently in effect.

12.4. Tariffing system during the LSB's exclusivity period provided for in Appendix I of Decree 62/90.

12.4.1. Revenues originated by the use of the STM and RTPN systems, shall belong to the corresponding licensees.
Without prejudice of the general principle stated in the above paragraph, STM and basic telephone service licensees may agree, in the interconnection agreements concluded, the economic compensations they deem necessary for the use of both networks.

12.4.2. Mobile service licensees may bill their subscribers for the provision of their service and what they agree with other services licensees in the interconnection agreements they conclude. Thus, the STM licensees shall be entitled to bill their subscribers, to order and account of the LSBs and the international service licensee (SPSI- Telecomunicaciones Internacionales de Argentina Telintar S.A.)). This criterium may be changed in the interconnection agreements.
STM licensees may not bill their subscribers, for the use of the systems belonging to the LSBs and the SPSI, amounts which differ from those of the basic telephone service.

12.5. For the purposes of the billing of services to roaming subscribers, current agreements in effect between operators and those which may exist between the Argentine Republic or other countries where such services become available, shall be respected.

Ministerio de Economía
y Obras y Servicios Públicos

SECTION No. 13. **Assignment of Rights.**

13.1. Licensees may not assign, transfer, encumber, or convey, in any manner, either in whole or in part, to third parties, the license or the rights deriving therefrom received through this Tender, without prior approval of the CNT.
After five years of the grant of the license, the CNT shall authorize such applications for transfer which they may receive, taking into account those elements or conditions of technical and business capacity which at the time justified the grant of the license.

13.2. Shares in the licensee company may be freely transferred with the exception of the following subsection, and the emission of shares to be sold by public offering is particularly permitted, however, the standards of incompatibility set by this specification shall be respected in all cases.

13.3. During the first three years to be counted from the date of the pe-award, share participations in the licensee company may not be transferred without the authorization of the CNT, which shall grant approval so long as the elements or conditions relating to technical and economic capacity which at the time justified the award of the grant, are not changed.

SECTION No. 14: **Penalties.**

14.1. Violation of the provisions of this Specification and/or of the engagements offered through the proposal and accepted by the CNT and/or the currently valid Resolutions made by same which are applicable to this type of services, and the breaches committed by a Licensee of the STM, either verified by the CNT ex-oficio and/or resulting from a complaint of a party or third parties, shall be subject to penalties, in accordance with their severity, by the guidelines of section 38 of Decree No. 1185/90 and in accordance with this specification by:

a) Warning
b) Fines
c) Total or partial revocation of the license or permit.

14.2. Warning or fine.

14.2.1. The following violations shall be subject to sanction by means of a warning or fine of up to the equivalent in pesos of three million (3,000,000) telephone pulses, as provided for by the CNT:

*Ministerio de Economía
y Obras y Servicios Públicos*

a) Total service interruption greater than twenty four (24) hours.

b) Partial service interruptions for periods greater than seven (7) consecutive days.

c) All violations of obligations imposed by the specification, in the license or in the applicable standards, which are not given a specific penalty.

14.2.2. The amount given in the previous subsection may be raised up to twelve million five hundred thousand (12,500,000) telephone pulses, in the event that non-compliance persists notwithstanding notification by the CNT or if the violation has grave social repercussions.

Within the maximum established, the CNT may apply fines for each day in which the non-compliance with the obligation persists.

14.3. Revocation of the license.

14.3.1 The following shall be sanctioned by revocation of the license granted:

a) The assignment or transfer to third parties of the license or the rights and obligations deriving from the same, without prior authorization by the CNT.

b) Any legal act which has the effect of encumbering the license granted.

c) The repeated occurrence of violations of the kind typified in section 14.2. on more than five (5) occasions annually, or persistent non-compliance for more than six (6) months counting from the first notification issued in this regard. This situation may give rise to partial revocation when the non-compliance takes place in a single location, or total revocation, in all other cases.

d) In the event that it is determined that, in accordance with the mechanism provided in section 25, the service is not in operation on the date, in the manner and quality contained in the proposal and agreed upon, independently of execution of the securety, the National Government may order partial revocation when the non-compliance takes place in a single location, or total revocation, when it takes place in more than one.

e) Receivership or Bankruptcy of the licensee company.

f) The dissolution or liquidation of the licensee company.

Ministerio de Economía
y Obras y Servicios Públicos

14.3.2. Procedure for revocation

14.3.2.1. The revocation based on the provisions of subsection 14.3.1.c) shall be preceded by formal notice to remedy the fault under sanction of revocation of the license, granting a period of five (5) days. In the event of subsequent non-compliance, the repetition of such notification shall not be necessary.

14.3.2.2. The revocation based on the provisions of subsections 14.3.1.a) and b) shall be preceded by a notice to correct the situation within a period of ten (10) days. In the event that such non-compliance is not corrected within that period, the provisions of subsection 14.3.3 shall be applicable forthwith.

14.3.2.3. The revocation based on the provisions of subsections 14.3.1.d), e) and f) shall result in the application forthwith of subsection 14.3.3.

14.3.3. Effects of the revocation

14.3.3.1. Total revocation

Once the penalty of total revocation of the license has been notified or in the event of abandonment of same the CNT may execute a management contract with an operator and take whatever other action it deems necessary to ensure adequate provision of service.
Except in case of receivership, where assets availment shall be made as provided for by the judge in charge, the former licensee's assets shall be transferred to the CNT in trust, so that it may auction them in the procedure for awarding a new license. In case of receivership the CNT shall require the judge in charge to follow the same procedure.
Except for the case of receivership, the former licencee company shall receive a maximum amount from the auction, consisting of the cost of assets excluding amortizations, expenses incurred by the CNT due to the revocation, taxes due, expenses due to the management contract, as well as the other necessary expenses incurred in providing the service.

14.3.3.2. Partial revocation

Once partial revocation of the license has been declared, for the causes provided in clauses c) and d) of section 14.3.1., a new bid for the provision of the service in the areas involved shall be invited, and the CNT -if deemed convenient, and if an installed infrstructure is available- shall be entitled to the procedures specified in the above section in that area.

Ministerio de Economía
y Obras y Servicios Públicos

14.4. Procedure for the application of penalties

Before application of penalties, the non-compliance shall be charged to the licensee company which shall be given a period of ten (10) working administrative days, to answer to the pertinent charges.

Once charges are answered or when the period in which they could have been answered expires, the CNT shall resolve the matter, without further substantiation and shall reliably give notification of the sanction applied.

The application of the penalty does not exempt the licensees from the fulfilment of his obligations. To such effect, upon notification of the sanction, notice shall be given to comply with the obligation, within a reasonable period to be determined, under sanction of application of additional sanctions. Non-compliance with such notification shall be deemed to aggravate the violation.

SECTION No. 15. **Jurisdiction.**

15.1. Any matter arising from the application or interpretation of the standards which govern the tender and/or the obligations contracted by participating in it or as a result of becoming a pre-awardee in the bid, and/or concerning the licenses and/or the provision of the services and/or, in general any question directly or indirectly related to the purpose and effects of the bid, shall be submitted to the National Courts of First Instance in Federal Administrative Litigations in the Federal Capital, with express exclusion of any other jurisdiction.

15.2. The submmission of a proposal implies express acceptance of the jurisdiction established in subsection 28.1.

Ministerio de Economía
y Obras y Servicios Públicos

TITLE III
CONDITIONS OF THE BID
Chapter I
Concerning the bidders

SECTION No. 16. **Requirements and exclusions**

16.1 Conditions

All corporations (sociedades anonimas) who fulfill the
requirements established in Chapter IV of this Title, may
present themselves in this bid, whether of domestic or
foreign capitalization, whether legally constituted
(chartered) or in process of (legal) formation in the
Argentine Republic, with exception of those specified in this
Specification and in Decrees Nos. 62/90, 677/90, 506/92 and
663/92.
A bidder shall be entitled to be awarded the bid and obtain a
licence in both operating areas.

16.2. Exclusions

Neither the licensees of the basic services who acquired the
former ENtel, nor the partners in their investment group,
whether related directly or indirectly through other
companies, may take part in this bid.

16.3. Participation in competing bands

Those who are partners in the company presenting the bid, as
well as their controlled or controlling companies, may not
participate in more than one proposal for the same operating
area.
This clause does not imply that a company which is a partner
in the company presenting the bid is forbidden to supply
technical assistance or equipment to other companies of which
it is not a partner.
The companies assisting the licensees of one band or which
are a part of their company, may not be a part of, nor assist
a company operating in a competitive band.

16.4. Limitation of operating areas

No legal person, nor their partners, may hold a license to
operate mobile telephony service, including SRMC, in more
than two operating areas in the Argentine Republic.

Chapter II
Procedure for the Bid
and Securities

SECTION No. 17. **Purchase of the Specification**

*Ministerio de Economía
y Obras y Servicios Públicos*

17.1. In order to submit a proposal in this bid, at least one of the members of the bidding consortium shall have purchased the present Specification of bases and conditions, under the conditions provided in the decree which approved it.

17.2. The purchase of a single Specification will enable the submission of proposals in both operating areas.

17.3. Submission of a proposal implies the acceptance of the terms of the present Specification and of the applicable standards established by the same.

17.4. Specification purchasers shall declare a legal domicile in the Capital City of the Argentine Republic for the purpose of receiving bidding notifications.

SECTION No. 18. **Enquiries**

Up to the date provided for in Appendix VI, the CNT will accept those enquiries which interested parties may make in written form, which shall be submitted to Sarmiento 151, fourth floor, office 435, Buenos Aires between 1:30 PM and 4:00 PM. The replies to such enquiries, together with the
    enquiries, shall be made known to all those purchasing the Specification.

SECTION NO. 19: **Maintenance of proposals.**

19.1. The proposals shall remain valid in the full extension of their terms up until the time of the pre-award of the license.

19.2. If such a pre-award contract is not executed within one hundred twenty (120) consecutive days, counted from the date of opening of the proposals, bidders may withdraw their proposals with ten (10) day's advance notice, without penalty. In the event that the first bidder in order of merit subject to evaluation or the pre-awardee, withdraws its bid, the next bidder's proposal in order of merit, shall become the one to be analyzed.

19.3. In the event that the proposal is withdrawn prior to the above term and without prior notice, the security posted shall be forfeited.

19.4. Submission of a single bid or the survival of a single bid due to the rejection or withdrawal of the others, shall not result in the suspension of the process of award of the bid.

SECTION No 20: **Reception and Opening of the bids**

Ministerio de Economía
y Obras y Servicios Públicos

20.1. The reception and opening of the Bids shall take place on the date provided for in Appendix VI, in a public session, in the Auditorium of the Subsecretaria de Comunicaciones (Subsecretariat of Communications), Sarmiento 151, fourth floor, Buenos Aires.

20.2. The participating authorities shall sign each of envelopes B and C to which reference is made in section 26.

20.3. After the signing of envelopes "B" and "C", they shall proceed to open envelope "A". Minutes shall be drawn up in which the bids received shall be listed, giving the name of the bidder, its integrating parties, operating area or areas proposed and any objections lodged by the bidders. The minutes shall be signed by the officer presiding over the meeting and the attendees who so wish. The originals of the bids are to be attached to the minutes as appendices to the same and shall be kept by the Ministerio de Economia y Servicios y Obras Públicas (Ministry of Economy and Public Works and Utilities) as reference documents.

20.4. One copy of each bid shall remain in the Subsecretaria de Comunicaciones (Subsecretariat of Communications), Sarmiento 151, fourth floor, Buenos Aires, where the bidders, their representatives and attorneys may take notice, take notes and at their own expense, make photocopies of the one evaluated for qualification or for the order of merit mentioned in section 21. The two remaining copies shall be sent to the Pre-award Commission for evaluation.

SECTION No. 21. **Order of merit and evaluation**

21.1. After bidders qualification is decided, qualified bidders shall be advised on the date for the opening of the bids.
Envelopes "C" of the qualified bidders shall be opened in a public session at the same place as mentioned in section 20.1. Following the opening of envelope "C", they shall proceed to the assignment of the point scores corresponding and relative to each proposal, thus establishing the order of merit of all of them.

21.2. The evaluation shall be made concerning the proposal with the highest score, for purposes of verifying that the data recorded on the control sheet for envelope "C" are in agreement with the record of the technical and economical proposal contained in envelope "B".
In the event that it is verified that the proposal agrees with the data entered in the control sheet, they shall

Ministerio de Economía
y Obras y Servicios Públicos

proceed to pre-award the license. In other circumstances, they shall proceed to disqualify the bidder and to evaluate the next offer in order of merit.

SECTION No. 22. **Participating Organizations**

The Subsecretaria de Comunicaciones (Subsecretariat of Communications) shall appoint a Pre-award Commission which shall function in the CNT area. The Pre-award Commission shall decide the qualification, evaluate the corresponding proposal and produce a report.
Said report shall be submitted to the Subsecretaria de Comunicaciones (Subsecretariat of Communications), which shall order its temporary approval adreferendum of the Ministry of Economy and Public Works and Utilities, to whom it shall be submitted for its definitive approval.
Once said Ministry has emitted its decision, the CNT shall notify the results to each of the bidders, by means of a registered letter and shall cause it to be displayed on the notice board of the CNT, Sarmiento 151, 4th. floor, Buenos Aires.

SECTION No. 23: **Objections.**

23.1. Bidders may file substantiated objections against the pre-award before the Pre-award Commission within the period provided for in Appendix VI, subject to the posting of a security deposit, which shall be returned in the event that the petition is honored, of two hundred thousand pesos ($ 200.000), in cash.

23.2. Such objections shall be submitted to the Pre-award Commission who shall produce a report to be submitted to the Subsecretaria de Comunicaciones (Undersecretariat of Communications) for resolution within the period provided for in Appendix VI.

SECTION No. 24: **Contracts for Pre-award of the licenses.**

24.1. The purchasers of the Specification shall be notified of the text of the pre-award contract, concerning which they may make suggestions and observations during the consultation period. The final text of the contract shall be notified to the interested parties after expiry of this period and before the period for submission of the proposals, such final text is to be presented in envelope "C", duly signed, within the terms of section 29.

24.2. Within the period provided for in Appendix VI, the Subsecretaria de Comunicacines (Undersecretariat of Communications) shall proceed to execute the respective pre-

Ministerio de Economía
y Obras y Servicios Públicos

award contracts for mobile telephony services in each of the operating areas given in Section No. 2.5.

24.3. If these pre-awards are made to companies in process of constitution, these shall be chartered within one hundred twenty (120) working days of notification of the preaward. This term may be extended one single time, by means of a decision of the CNT.

24.4. Until such a time as the company winning the pre-award of a license has been regularly chartered, each of the partners shall be jointly and severely liable, for the obligations assumed for the purposes of this tender.
Failure to obtain a regular charter on the part of the pre-awardee of a license within the term mentioned above, shall imply the automatic cancellation of its rights and the consequent loss of the securities.

24.5. In case a bidder wins in both operating areas, prior to the conclusion of the contract, it shall constitute an independent company for the provision of the service in the area other than the one in which the bidder company will provide the service.

SECTION No. 25: **Securities**

25.1. For the Proposal

25.1.1. To guarantee maintenance of the proposal until the time of pre-award, bidders shall post a bond equivalent to one million pesos ($ 1,000,000) whether the proposal is for one area, for the two areas or is a joint one. Proposals not meeting this requirement will be rejected.

25.1.2. Method of posting

Securities may be posted by means of any of the following methods: a) Security Bonds,
b) Bank Guarantees or
c) Deposit to the order of the CNT in an Argentine banking institution.

25.1.3. Return

Securities posted for proposal maintenance shall be officially and immediately returned, or freed of encumbrance, once the pre-award instrument is signed. Notwithstanding this, the CNT may return the securities once notification of the pre-award has taken place and even before the execution of the contract, to those bidders who so request it and who, in its sole judgment, are not likely to be accepted. The proposal security posted by those who turn out to be the pre-

Ministerio de Economía
y Obras y Servicios Públicos

award winners shall be returned to them or freed of
encumbrance, upon the posting of the pre-award security.

25.2. Concerning the pre-award and the licencse.

The bidders receiving the pre-award of the license, shall
post bond to secure the obligations and responsibilities to
be assumed, by means of the issue of a promissory note at
sight payable on demand upon simple expiry of the term for
payment, to order of the Govemment of the Argentine Nation,
guaranteed by their shareholders or by a first line bank
accepted by the Subsecretaria de Comunicaciones, for the
amount of two million pesos ($ 2.000,000) for each of the
mandatory service areas mentioned in Section 5.5 and as many
promisory notes as servise areas they offer to cover, for the
amount of five pesos ($5) per inhabitant, with a minimun of a
million pesos (1.000.000) per area.
This securety is to be posted at the time of execution of the
contract and shall include an appointment, as trustee, of a
first line International Bank acceptable to both parties, who
shall be instructed to deliver the bond(s) to the National
Government or to the pre-award winner, based on whether or
not the service is operational in the area to be covered on
the date of maturity, within the conditions proposed. The
functioning of the system within the proposed terms shall be
established by the CNT based on the findings of a
telecommunications technical consultant proposed by the pre-
award winner, among those mentioned in appendix IV.
The security documentation shall be exempt from stamp tax.

Chapter III
Concerning the Bids

SECTION No. 26. **Presentation format**

26.1. Proposals shall be addressed to the Subsecretaría de
Comunicaciones-Comisión Nacional de Telecomunicaciones.
Intrested parties are entitled to submit individual proposals
for one or both operating areas. They are also entitled to
submit a joint proposal specifying the point ratings
separately for each area in order to compare them with the
other proposals. The license will be awarded to a joint
proposal only if it is assigned the highest score in both
areas.

26.2. All of the documents detailed in Sections 27 and 28
shall be presented in compliance with the format established
in Appendix VI of the present Specification.

26.3. Proposals are to be presented in quadruplicate, in
three sealed envelopes, marked ''A", "B" and "C', contained
in an envelope on which shall be written the company's name

Ministerio de Economía
y Obras y Servicios Públicos

or identity of the person submitting the proposal, together with the legend "Ministerio de Economía y Obras y Servicios Públicos — Ministerio de Economía y Obras y Servicios Públicos — Secretaria de Obras Públicas y Comunicaciones — Subsecretaria de Comunicaciones — Licitación para la provisión del Servicio de Telefonia Móvil" (Ministry of Economy and Public Works and Utilities of the Argentine Republic — Secretariat of Public Works and Communications — Undersecretariat of Communications — Tender for the provision of mobile telephony service).

26.4. The bids and documentation enclosed shall be drawn up in the Spanish language or in a foreign language, so long as the respective translation into Spanish is enclosed, effected by a matriculated national certified translator. Documentation which has been issued by official organizations or notaries of foreign countries shall be presented legalized by the Ministerio de Relaciones Exteriores y Culto de la Argentina (Ministry of Foreign Affairs and Religion of Argentina), through its diplomatic legations in the country of origin.
Documents originated in signatory countries of the Hague Convention may be legallized according to said Convention.

26.5. All of the sheets of the proposal shall be consecutively numbered in the upper right hand corner and signed by the legal representative or attorney of the bidder.

26.6. Simple copies of all documents which are to be presented in original form, in accordance with the requirements given in Sections Nos. 25, 29, 30 and 31, shall be acceptable for the duplicate, triplicate and quadruplicate of each envelope.

SECTION No. 27: **Contents of envelope "A".**

The envelope "A" shall include the following documentation, authenticated by a notary public, when enclosing photocopies of probatory documents.

27.1. General index, with indication of the pages on which the proposal is to be developed.

27.2. The following information concerning the bidder:

— Legal Name.
— Type of corporation and mention of whether it is chartered or in process of formation.
— Legal domicile.
— Special domicile constituted for purposes of this tender in the Federal Capital.

Ministerio de Economía
y Obras y Servicios Públicos

— Names, surnames and identification document number of the holders of powers of attorney who are signatories of the proposal.

—The share participations of the partners forming part of the bidder's company shall be explicitly indicated in the proposal.

These, as well as the other aspects included in the documentation presented at the time of the bid, may not be amended, without the participation of the CNT, up to the moment that the present bid is awarded. In the event that this requirement is not fulfilled, the CNT may, according to the severity of the non-compliance committed, reject the proposal.

27.3. The following Legal documentation of the bidder:

27.3.1 Corporations which are legally constituted (chartered/incorporated), shall prove their constitution in the manner prescribed by law and other standards regulating the matter.

27.3.2 Corporations in process of formation, shall attach a certified copy of the partnership agreement or of the minutes of the constitutive meeting signed in both cases by all of the partners. Each of the partners who are legal persons shall document their constitution in the manner prescribed by law and other regulations applicable to the matter.

27.3.3. Documents proving the legal representation of attorneys signing the proposal.

27.4 Background and references of the bidder in accordance with the stipulations of Section 30.

27.5. The following documentation evidencing the business-financial capacity of the bidder:

A) Photocopies of the balance sheets for the last two (2) fiscal years, signed by a responsible auditor and certified by Notary Public. In the event that a partner were to have been in business a shorter time, the requirement shall be for whatever balance sheet is available and an actuarial certification of the equity involved.

B) Details of the amount which each partner has committed to invest in the project and the ratio between said amount and his net worth. For purposes of prequalification, only those proposals which are consistent with the economic capacity represented by this parameter, shall be taken into account.

Ministerio de Economía
y Obras y Servicios Públicos

In the event that the last audited balance sheet was previous to 31 December 1991, a statement from the auditors shall be presented indicating that, to the best of their knowledge and belief, from the date of the last balance sheet to that of the proposal, neither the net worth nor the solvency have undergone any substantial negative variation.

27.6. Guarantee for the proposal in accordance with the requirements of Section No. 25.1.

27.7. Specification of the present bid.

27.8. Receipt for the purchase of the Specification for the present bid.

27.9. Indication of the operating area(s) for which the proposal is being submitted.

SECTION No. 28: **Contents of Envelope "B".**

This shall contain (separately for each one of the operating areas for which the bidders are tendering):

28.1. General index, with indication of the pages on which the proposal is to be developed.

28.2. General descriptive summary of the project proposed to be developed for operation of the service, in accordance with the requirements of this Specification.

28.3. Detail of the service areas which they propose to cover, as well as those areas which, at minimum, are required in section 5.5., giving the dates on which it is proposed to provide such coverage of service area. Each of these areas shall be analysed in the tecnical plan per requirement given below.

28.4. Design of the network planned.

28.4.1. A plan covering the design of the planned network, annually, for the first five (5) years of service operation shall be included, which shall be consistent with the quality of service and the assumptions under the present Specification regarding the market share percent and the traffic percent per subscriber.
Such a plan shall contain a forecast of the coverage of the service areas required in section 5.5., as well as those additional ones which the bidders have included in their proposal.
To evaluate the consistency of the technical proposal, the guidelines of quality, power and blocking, contained in Resolution 498 SC/87 shall be taken into consideration.

*Ministerio de Economía
y Obras y Servicios Públicos*

28.4.2. This plan shall give a detail of: cellular infrastructure, control and switching centers (CCM and CCR), points of interconnection with the public telephone network, links between cell sites and switching and control centers and location of base stations.

28.4.3. The design shall have attached the technical calculations in order to permit verification that the coverage and the traffic and switching capacity required for the population coverage proposed, will be satisfied, as well as a certification by the firm or professional making the radio coverage calculations.

28.5. Construction work plans outlining what is to be accomplished before start-up in each service area, giving the latest date on which service is to begin, in each of these.

28.6. Design of the computer center, giving the features of the hardware and software to be installed, which will provide for service billing in accordance with the stipulations of Sections Nos. 9 and 12. The individual subsystems shall be specified for activating and deactivating subscribers, for follow-up of subscriber complaints, etc., as well as everything pertaining to the quality of the customer service to be provided.

28.7. Investment plans for five (5) years of service, to be compatible with the plan presented and the assumptions regarding market penetration (including the expansion planned in the service areas). The investment plan shall not be less than fifteen million United States dollars (U$S 15,000,000) per operational area in the first three (3) years counting from the date of the pre-award.

28.8. Documentation evidencing the business-financial capacity of the bidder, in accordance with the following detail:

A) Cash flow and total to be invested in the project, in United States dollars, for each one of the areas for which a proposal is being submitted, including the total investment in infrastructure and operating costs until the point of self-financing is achieved.

B) Statement of sources and application of funds corresponding to the last period for which accounting records are being submitted.

C) Bank references and other information of a financial nature such as to facilitate evaluation, including details

Ministerio de Economía
y Obras y Servicios Públicos

concerning source of financing for the project, together with appropriate certification by same indicating their willingness to do so and the conditions to be met in order to obtain the indicated financing.

D) Table giving projected profit and loss, by years, for five (5) years of the project, together with the projected cash flow, showing the internal rate of return projected.

The information to be included, listed in items B) and C) above, shall be that of the corporation and its partners, or the legal persons which directly or indirectly control them.

SECTION No. 29. **Contents of envelope "C"**

Envelope "C" shall include:

29.1. The control sheet, whose form shall be delivered by the CNT to the purchasers of the Specification, filled out with the data on coverage and time of operation contemplated in the proposal.

29.2. The final text of the pre-award contract executed by the legal representative of the bidding company.

29.3. Total amount of the securities established in section 25.2., in order to make it possible to verify the compliance with the same related to the proportion of securities as compared to net worth.

Capter IV
Requirements for qualification

SECTION No. 30. **Bidder Background and References**

30.1. It shall be a condition for qualification of the proposals submitted that operators of mobile cellular telephony services have a minimum participation of twenty percent (20%) in the capital stock of the bidder.
For purposes of this clause, operator shall be taken to mean the company or companies who document their experience in the provision of mobile cellular telephony services and/or the development of technology applicable to that service. Companies having one or both of the mentioned kinds of experience, shall jointly possess twenty percent (20%) of the capital stock, although those having operating experience in the provision of the service may not have a share which is less than fifteen percent (15%).

30.2. For purposes of qualification, operators shall have one of the following backgrounds as a minimum:

*Ministerio de Economía
y Obras y Servicios Públicos*

a) In addition to providing mobile telephony, provide wire line telephony services.
b) Have a number of mobile telephony subscribers not less than one hundred and fifty thousand (150,000).
c) Be companies providing the service in countries members of the ALADI (Latin-American Association), proving permanence in the use of their license, and having references of provision of good service issued by the governments of the countries in which they are being provided.

Only the background corresponding to companies providing the service either directly or by means of a company providing the service, shall be computed as operators background. A company shall be considered as having an operator background when the company has not less than thirty (30) per cent of the shares in the company which provides the service.
If the company which is a partner of a bidder is, at the same time, directly or indirectly controlled by a holding company, the experience and backgrpund of other companies directly or indirectly controlled by that holding company, with no less than the fifty one percent (51%) of the capital stock, shall be computed for the purpose of this section and at request of the interested party.

30.3. Bidders shall substantiate such experience by means of:

A) An affidavit including a list of the cellular services which they provide or supply, either of themselves or through one of the legal persons mentioned in the previous
     paragraph, giving in each case: areas and population covered, number of subscribers, years in operation of the service, shareholding in the service provider (either of the
     bidder and/or of the operators which constitute it) and any other data enabling evaluation of the bidder´s experience.

B) Information regarding the type and size of the cellular switching exchange(s), number of cell sites and number of channels for each one of the services provided or operated, shall be declared by way of reference.

SECTION No. 31: **Business-financial capacity of the bidder.**

For a proposal to be qualified, the members of the proposing company shall have a joint net worth, of not less than one hundred million United States dollars (U$S 100,000,000); all other cases the proposal shall be rejected forthwith. In order to be awarded two areas a bidder shall document a business and financial capacity which is double that of the parameters indicated.

*Ministerio de Economía
y Obras y Servicios Públicos*

The net worth of those partners who have less than fifteen (15) percent participation in the company submitting the bid, shall not be entered into the computation.
For the purpose of this section, the information being considered shall be, at request of the interested party, the one corresponding to the company which directly or indirectly controls the partner of the bidding company, or, when appropiate, the one corresponding to all the companies directly or indirectly controlled by that controlling company, with more than fifty one (51) percent of the capital stock.

Ministerio de Economía
y Obras y Servicios Públicos

Chapter V
Evaluation of proposals and selection of pre-
awardees

SECTION No. 32. **Qualification of the Bidders**

32.1. The qualification of the bidders shall be undertaken by
means of a study of the documents included in envelope "A".
Proposals which do not fulfill the requirements given in
sections Nos. 25, 27, 30 and 31, shall be rejected forthwith
and envelopes "B" and "C" returned to the corresponding
bidder.

32.2. If the proposal fulfills the minimum requirements of
chapter IV of this title, it shall be considered as qualified
and shall warrant analysis of the proposal of envelope "C".

32.3. If during evaluation of envelope "A" it were to appear
that all or part of the documentation presented were
considered lacking in sufficient clarity, the Pre-award
Commission may request all clarifications, which do not imply
changes in the proposals submitted by a bidder, that it deems
necessary.

SECTION No. 33. **Selection of the pre-awardee**

33.1. Those proposals which do not enclose all of the
documentation required in Envelope "B", shall be rejected
forthwith. Notwithstanding, if in the judgment of the Pre-
Award Commission the only clarifications necessary were those
which do not imply changes in the proposal presented by a
given bidder, it may request those it deems necessary.

33.2. Bids which do not comply with the requirements of the
present Specification or which contain clauses which oppose
them, shall be rejected.

33.3. Evaluation of coverage shall be made using the CAREY
90/90 system.
The selection of the pre-award winner shall follow upon
evaluation of Envelope "C" and its consistency with the
technical/economical project of envelope "B".
This evaluation (between 0 and 100 points) shall be made by
means of the application of the following algorithm:

$$C = P \times 0.30 + G \times 0.50 + V \times 0.20$$

Where:

C: Is the final point score of each bidder.
P: Is the total population coverage of each proposal at the
end of five years counted   from the day of the pre-award.

Ministerio de Economía
y Obras y Servicios Públicos

G: Geographic coverage of each proposal at the end of five years counted from the day of the pre-award.
V: The time in which service shall be provided in the mandatory service areas of section 5.5.

33.5. Total population coverage

33.4.1. For purposes of evaluating the population coverage, the following guidelines shall be taken into account:

a) The population which arises for each city in Appendix V of the present Specification, shall be taken as the base.
b) Point scores shall only be given to coverages of 100% of the political limit of each city.

33.4.2. For purposes of comparison of the various proposals, the one with the greatest coverage shall be assigned a total of 100 points and the remaining ones shall be assigned point scores on a proportional basis.

33.4.3. The population coverage shall have a relative point score of thirty percent (30%) of the total.

33.5. Geographic coverage

33.5.1. For purposes of evaluation of geographic coverage, the following guidelines shall be taken into account:

a) The population covered by service areas outside of the political limits of the cities mentioned in section 5.5, shall be assigned a point score.
b) The length of roads covered in lineal kilometers, shall be assigned a point score, in the following corridors: Buenos Aires-Rosario (Ruta Nacional 9); Rosario-Cordoba (Ruta Nacional 9); Buenos Aires-Mendoza (Rutas Nacionales 7 y 8); Buenos Aires-Bahia Blanca (Ruta Nacional 3); Rosario-Ceres (Ruta Nacional 34); y Buenos Aires-Mar del Plata (Ruta Nacional 2).
c) The figures mentioned in the previous clauses shall be reduced by 1.666% per month, in accordance with the temporary commitment of start up, in such a way that coverage during the first month counted from the pre-award shall correspond to 100% of the numbers and that of coverage during the last month of the fifth year shall be of 1.666%.

33.5.2. Evaluation of geographic coverage shall be made using the following algorithm:

$$G = Gp + Gr$$

Where:

Ministerio de Economía
y Obras y Servicios Públicos

Gp is equal to geographic population coverage and shall assign a value of 60 points to the best proposal, and a proportional value to the remaining ones.

Gr is equal to coverage of the mandatory roadways and will assign 60 points to the best proposal and a proportional value to the remainder.

33.5.3. The geographic coverage shall have a relative point score of fifty percent (50%) of the total.

33.6. <u>Time to start-up in the mandatory service areas</u>

33.6.1. For purposes of evaluation of the time in which service shall be provided in the mandatory areas, the following guidelines shall be followed:

a) The months to start up estimated in section 5.5 shall be taken as the base.
b) Points will be given according to the months to start up less than the estimate given in section 5.5, on a city by city basis.
c) The best proposal shall be assigned a total of 100 points per city, the point score for the remainder being assigned on a proportional basis.
d) The point score for all of the cities made by each of the bidders, shall be averaged, the highest average shall be assigned 100 points, the remainder being assigned a proportional point score.

33.6.2. The speed with which service is started up shall have a relative point score of twenty (20%) of the total.

SECTION 34. **Estimates of market share percent and traffic**

3.4. For coverage calculation purposes, the following assumptions on the market share percent on covered population, shown in percentages of total population, shall be taken into account (pops):

POPS	YEAR 1	YEAR 2	YEAR 3	YEAR 4	YEAR 5
1,000,000 or over	0.1	0.15	0.2	0.35	0.5
400,000 to 1,000,000	0.06	0.1	0.15	0.25	0.35
200,000 to 400,000	0.05	0.07	0.1	0.15	0.25

Ministerio de Economía
y Obras y Servicios Públicos

34.2. Considering the exclusivity period established and taking into account the future operation in competition with a coprovider of the service. for the purpose of the tecnical project. on the basis of the figures above mentioned, the total incremental market share percentage of the year shall be reduced fifty (50) percent as from the third year after pre-award.

34.3. For the purpose of developing the technical plan, an average monthly traffic per subscriber of two hundred (200) minutes/subscriber-month, shall be considered.

Ministerio de Economía
y Obras y Servicios Públicos

APPENDIX II

BOUNDARIES AND DISCRIPTION OF THE OPERATING AREAS

AREA I

Operating Area I is comprised of the Provinces of Entre Rios, Corrientes, Misiones, Córdoba, Santiago del Estero, Chaco, Formosa, Catamarca, La Rioja, Tucumán Salta and Jujuy and the Province of Santa Fe with the exclusion of the land boundaries of the Departments of Constitución, San Lorenzo and Rosario.

AREA III

Operating Area III is comprised of the Provinces of San Juan, San Luis, Mendoza, La Pampa, Neuquén, Río Negro, Chubut and Tierra del Fuego, Antártida e Islas del Atlántico Sur, the Departments of Constitución, San Lorenzo and Rosario, and the Province of Buenos Aires, with the exclusion of Operating Area II, also called "AMBA and its extensions".

AREA II

Operating Area II is comprised of:

The Multiple Area of Buenos Aires (AMBA), The Multiple Area of La Plata and the La Plata-Buenos Aires corridor.

The Municipalities of Tigre, Escobar, Campana, Zarate, Pilar, Luján, General Rodriguez, General Sarmiento, Moreno, Merlo and La Matanza of the Province of Buenos Aires.

The area inbetween the River Plate's coast and the lines that join the following geografic coordinates:

LATITUDE (S)	LONGITUDE (O)
34° 24′ 00″	58° 26′ 34″
34° 25′ 59″	58° 20′ 21″
34° 29′ 56″	58° 13′ 44″
34° 38′ 07″	57° 57′ 28″
34° 42′ 24″	57° 49′ 08″
34° 44′ 15″	57° 45′ 59″
34° 47′ 53″	57° 41′ 51″
34° 53′ 17″	57° 40′ 20″

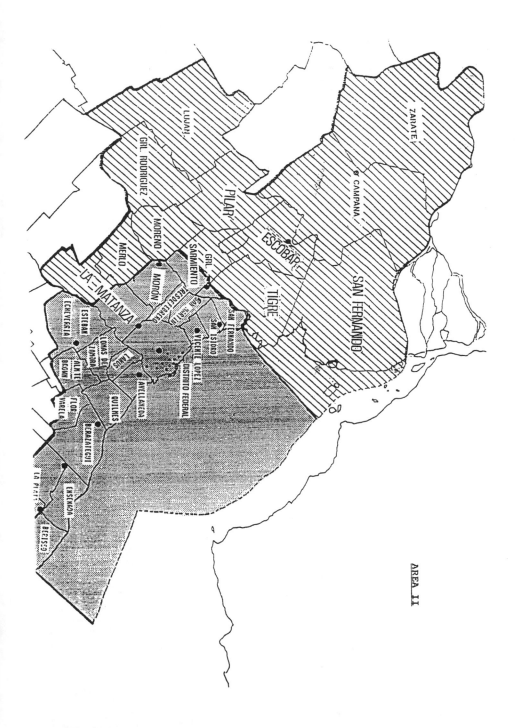

Ministerio de Economía
y Obras y Servicios Públicos

APPENDIX III

LIST OF TECHNICAL CONSULTANTS

Coopers & Lybrand (UK)
Booz Allen & Hamilton (EEUU)
Deloitte & Touche (EEUU)
Price Waterhouse (Canada)
Arthur D. Little (EEUU)

Ministerio de Economía
y Obras y Servicios Públicos

APPENDIX VI

TERMS OF THE PUBLIC BID FOR THE PROVISION OF
MOBILE TELEPHONY SERVICES

1. The CNT will accept the enquiries which the interested parties may make in written form, until thirty (30) working days prior to the date provided for the opening of the bids.

2. The reception and opening of the bids shall take place on day sixty (60), counted from the date the specification begins to be sold. This term can be extended by the Pre-award Commission for thirty days on substantiated request of more than one bidder.

3. Bidders may file substantiated objections against the pre-award before the Pre-award Commission in a term not in excess of five (5) working days counted from the date of the notice.

4. Within ten (10) working days from the expiry of the term in which objections may be filed, the Pre-award Commission shall produce a report to be submitted to the Subsecretaria de Comunicaciones for resolution.

5. The Subsecretaria de Comunicaciones shall proceed to excecute the respective pre-award contracts for mobile telephony services in each of the operating areas, within thirty days from the expiry of the term to file objections, in case no one is filed, or from the resolution of those which may have been filed.

APPENDIX B
▼▼▼

INVITATION TO APPLY FOR A LICENSE TO PROVIDE A NATIONAL CELLULAR TELEPHONY SERVICE ISSUED BY THE SOUTH AFRICAN MINISTER OF TRANSPORT AND OF POSTS AND TELECOMMUNICATIONS

Invitation to Apply

for a Licence to

Provide a National Cellular

Radio Telephony Service

IMPORTANT NOTE TO RECIPIENTS

This document constitutes an invitation to apply for a licence to construct, operate and maintain a nationwide cellular radio telephony service on a non-exclusive basis.

The right is reserved to alter, correct, suspend or abandon the process of selection of a licensee at any stage prior to the issuing of a licence.

In the document some information and guidelines are given in order to assist Applicants. Although care has been taken to give correct information, it cannot be guaranteed. Responsibility is therefore disclaimed in respect of the accuracy or otherwise of the information. Applicants should satisfy themselves on the correctness of the information, and in so far as may be necessary, consult their own legal and other professional advisers.

Release of this document in no way binds the Minister or the Postmaster General to follow a particular process with regard to the issuing of the licences.

....

DR P.J. WELGEMOED

MINISTER OF TRANSPORT AND OF

POSTS AND TELECOMMUNICATIONS

<u>*INDEX*</u>

CHAPTER 1

Interpretation

CHAPTER 2

Introduction

CHAPTER 3

Invitation to Apply for a Licence

CHAPTER 4

Information to be Given

CHAPTER 5

Selection Criteria

CHAPTER 1

INTERPRETATION

Definitions

1.   *In this document and in any application unless the context otherwise indicates -*

  (a)   *"AMPS" shall mean the Advanced Mobile Phone System and "D-AMPS" shall mean the digital version thereof.*

  (b)   *"Applicant" shall mean -*

    (i)   *a company formally applying for a licence; or*

    (ii)   *a company to be formed on whose behalf such an application is made (as the case may be)*

  to construct and operate a cellular radio telephony service;

  (c)   *"application" shall mean the application submitted by the Applicant for the authority and licence referred to in paragraph 9;*

(d)   *"associated company" shall mean a company which in the opinion of the Minister is related to another company in such a way that there is (directly or indirectly) a substantial identity of interest between the companies, their subsidiaries, controlling or holding companies, their shareholders (a nominee and his principal is to be regarded as the same person) or the shareholders of their controlling, holding or subsidiary companies and shall without limiting the discretion of the Minister include -*

(i)   *the controlling company of the company concerned;*

(ii)   *a subsidiary company of the company concerned;*

(iii)   *a subsidiary company of the controlling and subsidiary companies referred to in (i) and (ii) respectively;*

(iv)   *a company having a shareholder, director, executive manager who is also a shareholder, director, executive manager of the company concerned;*

(v)   *an associated company of the associated company referred to in (i) to (iv);*

(e)   *"Base Station" and "BSS" shall mean a base station system comprising a Base Transceiver Station (BTS) controlled by a Base Station Controller (BSC) within the meaning of the GSM Recommendations;*

(f)   "Basic Services" shall mean telecommunications services defined in the GSM Recommendations 02.02 and 02.03;

(g)   "BSC" shall mean Base Station Controller;

(h)   "BTS" shall mean a Base Transceiver Station;

(i)   "Cellular 2", "Cellular 3", "C2" and "C3" shall mean the licences respectively referred to in chapter 2, paragraph 6;

(j)   "CEPT" shall mean the Conference of European Posts and Telecommunications;

(k)   "counter trade" shall mean business which a foreign principal undertakes to place on a South African company as a direct result of having made sales of imported equipment (the nature of which shall be similar to that of the imported equipment, although it need not be the identical product) in South Africa;

(l)   "EMC" shall mean Electro Magnetic Compatibility;

(m)   *"ETSI" shall mean European Telecommunications Standards Institute;*

(n)   *"fixed network" shall mean any telecommunications network which links two fixed points and shall include the public switched network ("PSTN") leased lines, data lines and data network;*

(o)   *"GSM" shall mean the pan-European digital telephony standard for cellular radio communications known as Global System Mobile since 1990, operating in the 900 Mhz frequency band;*

(p)   *"GSM Recommendations" shall mean the European standards issued by ETSI for GSM;*

(q)   *"IFRB" shall mean the International Frequency Registration Board;*

(r)   *"ISDN" shall mean Integrated Services Digital Network;*

(s)   *"ITU" shall mean the International Telecommunications Union;*

(t)   *"licence" shall mean both -*

    (a)   *the authority given by the Minister to construct and operate a cellular radio telephony service in terms of section 90A(1) of the Post Office Act, 1958; and*

    (b)   *the licence issued by the Postmaster General in terms of section 7 of the Radio Act, 1952, conferring the right to use or cause any person in his employ or under his control to use a station;*

(u)   *"licensee", "licence holder", "operator", shall all mean an Applicant who has been granted a licence to contruct, operate and maintain a cellular radio telecommunications system;*

(v)   *"Minister" shall mean the Minister of Transport and of Posts and Telecommunications or, in the case of a change of portfolio description, the Minister to whom the administration of section 90A(1) of the Post Office Act has been assigned, or his delegate;*

(w)   *"mobile radio services" shall mean bearer services, telecommunications services and supplementary services within the meaning of the GSM Recommendations;*

(x) *"Mobile Stations" or "MS" shall mean a mobile radio station within the meaning of the GSM Recommendations;*

(y) *"MSC" shall mean a telephone exchange for cellular telephony and referred to as a Mobile Switching Centre within the meaning of the GSM Recommendations;*

(z) *"Net Revenue" shall mean the total income derived from the sale of cellular services less all returns, trade discounts, or funds collected on behalf of others;*

(aa) *"NMT" shall mean the Nordic Mobile Telephone System;*

(bb) *"PSTN" shall mean the Public Switched Telephone Network operated by Telkom;*

(cc) *"roaming" shall mean the use of a mobile station on a network other than the network to which a user subscribes;*

(dd) *"service" shall mean the cellular radio telephony service refered to in paragraph 9;*

(ee) *"service provider" shall mean a person (natural or corporate) offering digital cellular radio services as a retailer;*

(ff)   "TACS" shall mean the Total Access Communication System;

(gg)   "Telkom" shall mean Telkom SA Limited.

Captions

2.   Captions are used for convenience only and shall not be taken into consideration for interpretation purposes.

Number

3.   Unless the context otherwise indicates, words in the singular number include the plural and words in the plural number include the singular.

CHAPTER 2

INTRODUCTION

Minister's announcement

4.   The Minister recently announced his intention to grant licences to construct, operate and maintain two nationwide mobile cellular radio systems *(annexure "A").*

Frequencies

5.   Frequencies have recently been made available in the 900 MHz band and the cellular systems may be based on the frequencies within the bands 890-915 MHz *(mobile transmit)* and 935-960 MHz *(base transmit)* or such other frequencies below the 960 MHz band as the Minister deems appropriate. In this connection some further frequencies may be available in some areas either now or in the future and Applicants are encouraged to suggest frequency bands which they consider appropriate to their systems.

Licences

6.   (1)   The two new licences will be referred to as Cellular 2 or C2 and Cellular 3 or C3.

(2)   *C2 will be issued to a company in which Telkom holds shares of not more than 50%.*

(3)   *The C2-company must nevertheless submit a full licence application on or before 30 June 1993.*

(4)   *This document constitutes an invitation to apply for the C3 licence.*

(5)   *The C2 and C3 licence holder shall operate under precisely the same terms of licence and at arms length to Telkom.*

(6)   *Both licensees shall be given the same date upon which the network may be switched on.*

(7)   *The switch on date referred to in sub-paragraph (6), shall allow adequate time for both licensees to complete the construction of their respective networks.*

Choice of technology

7.   *All Applicants may submit proposals for the technology and standards of their choice. As GSM is widely acclaimed in South Africa any other technology proposed should include a full discussion comparing it with GSM.*

Procedure

8. *A fair procedure will be followed in selecting the C3 licensee. The Minister and Postmaster General will in their consideration of the licence be advised by experts.*

CHAPTER 3

INVITATION TO APPLY FOR A LICENCE

Invitation

9.   Companies incorporated in terms of the Companies Act, 1973, or persons acting on behalf of such companies to be formed are herewith invited to apply -

(a)   to the Minister in terms of section 90 A (1) of the Post Office Act, 1958 for authority to construct, operate and maintain a national cellular radio telephony service on a non-exclusive basis; and

(b)   to the Postmaster General in terms of section 7 of the Radio Act, 1952 for a licence(s) to be issued to use radio stations in the appropriate frequency bands in respect of the rendering of the service referred to in paragraph (a).

Disqualifications

10.   *(1)*   *The following companies are disqualified from becoming or being a licence holder :*

       (a)   *A company not controlled by South African citizens ordinarily resident in South Africa;*

       (b)   *a company which -*

              (i)   *holds shares in another prospective licence holder or associate company of such licence holder; or*

             (ii)   *has a financial interest in such licence holder;*

       (c)   *a company of which any shareholder, member of its top management or any director -*

              (i)   *holds shares giving such shareholder, member of top management or director (jointly or severally) a financial interest of more than 5% in another licence holder or any associate company of such licence holder;*

        (ii)   has any financial interest of more than 5% in such licence holder;

    (d)   a company which has one or more shareholders acting as nominee or agent for another person;

    (e)   a company which is a manufacturer or supplier of telecommunications equipment or a company which is an associated company of a company that manufactures or supplies telecommunications equipment where such association may have a material impact on competition.

(2)   The Minister shall in his sole discretion determine whether a company is disqualified from being a licence holder in terms of subparagraph (1).

Format of applications

11. (1)   Applications must be -

    (a)   typed or printed and provide the information prescribed in chapter 4 in the same sequence and with the same headings, or otherwise be accompanied with suitable cross-references;

(b)   *written in the English language and may not comprise more than 200 A4 pages (single spacing) including all the annexures, but excluding the summary;*

(c)   *accompanied by a summary of not more than 25 pages.*

(2)   *Should the Applicant offer alternative technologies then, notwithstanding the provisions of sub-paragraph (1)(b), each alternative technology may occupy additional pages.*

(3)   *Should an Applicant consider it necessary to provide additional information which cannot be suitably included under the subjects specified in chapter 4, the Applicant may add further subjects.*

Number of copies

12.   *One original and 15 copies of the application must be submitted.*

Duplication to be avoided

13.   *Applicants should avoid duplication of information.   A suitable cross-reference will, however, be helpful.*

Submission Address and Application fee

14. (1) *Sealed applications must be forwarded by registered post or delivered to -*

> *The Postmaster General*
> *Department of Posts and Telecommunications*
> *8th Floor Old Mutual Building*
> *Andries Street*
> *Private Bag X860*
> *PRETORIA 0001*

(2) *An application fee of R50 000 (fifty thousand Rand) must be paid simultaneously with the submission of the application.*

(3) *Applications may be withdrawn in writing before the closing date by letter or telefax.*

Deadline for Submission of Applications

15. (1) *The closing date for the submission of applications will be 12 noon on 1 June 1993 or such other date as the Minister may determine.*

(2) *Any application received after the closing date shall be disqualified.*

Costs Incurred in Making Application

16. Applicants shall bear their own costs regarding their applications as well as any other costs incurred before or after the granting of a licence.

Deadline for Granting the Licence and Validity Term

17. (1) It is intended to grant the licences by 30 June 1993 or as soon thereafter as possible.

(2) Every application must state that it is irrevocably open for acceptance until 30 September 1993.

Enquiries or Comments

18. (1) Enquiries or comments regarding this invitation may be directed in writing to the address given in paragraph 14 above within the first two weeks after its issue.

(2) In so far as such enquiries or comments are answered, the questions and answers will be sent to all those who have given notice of their intention to submit an application.

(3)   *The identity of the Applicant asking the question or making comments will not be disclosed.*

(4)   *It may be regarded as an irregularity if an enquiry or comment is made in any other way.*

Clarification of Application

19. (1)   *Should the Minister require clarification from an Applicant he may invite such Applicant for discussions.*

(2)   *Alternatively, the Minister may require further documents or other clarification in order to facilitate the decision making process.*

(3)   *The Minister may request the Applicant to submit such information subsequent to the closing date for applications.*

Consent to Publication

20. *Each Applicant must in his application consent to the publication of his name as an Applicant as well as the composition of his shareholding, should a licence be granted to him.*

Notice of award

21.   All Applicants (successful and unsuccessful) will be informed in writing of the decisions of the Minister and the Postmaster General.

Non-compliance with Provisions

22.   Failure to comply with the provisions of this document, may lead to the disqualification of an application.

Draft Structure of Licence, Scope of Licence and Fees

23.   (1)   A draft of the structure of the operating licence and the radio licence together with typical radio application forms are contained in Annexure B.

      (2)   Non-compliance with licence conditions may inter alia lead to penalties being imposed, suspension and cancellation of the licence.

      (3)   The licence will be transferable with the written consent of the Minister and the Postmaster General.

      (4)   The term of the licence shall be 15 years.

(5)   It is contemplated that interconnection of the networks of the C2 and C3 licensees to Telkom's fixed networks will be arranged by way of an agreement to be negotiated and concluded by the C2 and C3 licensees and Telkom.

(6)   More particulars regarding the agreement are contained in annexure "C".

Licence Fees

24. (1)   Both licensees shall pay -

(a)   An initial basic cellular licence fee of R100 000 000 (one hundred million Rand), payable

(i)   prior to the commencement of commercial operations; or

(ii)   by way of instalments agreed upon with the Regulator.

(b)   An ongoing annual licence fee of 5% of the net revenue of the licensee concerned, payable within 14 days after the end of each licence year.

(2)   *The licence fees shall be payable to the official designated by the Minister to receive such payment.*

(3)   *Both licensees shall pay the following radio fees to the Postmaster General :*

(a)   *a basic annual licence fee of R5 000 000 (five million Rand);*

(b)   *a fee of R20 000 (twenty thousand Rand) annual licence fee for each 200 KHz channel granted to the licensee concerned.*

CHAPTER 4

INFORMATION TO BE GIVEN

Information about the Applicant

25.   *The following information must be given in the application :*

(a)   *The name, address and registration number (if available) of the Applicant, the names of its directors, principal executives and main objective.*

(b)   *If applicable and in so far as it is available, the Annual Reports of the previous three years of -*

(i)   *the Applicant; and*

(ii)   *each shareholder being a company, provided that such shareholders hold more than 5% of the shares in the Applicant.*

(c)   *The names and addresses of all shareholders or prospective shareholders of the Applicant and of all companies having a direct or indirect financial interest in the Applicant in a table together with their shareholding and all future rights.*

(d)   Whether the Applicant, its shareholders or any associated companies of the Applicant are manufacturers or suppliers of telecommunication equipment or have any financial interest in such a manufacturer or supplier.

(e)   Full particulars of the experience and expertise of the Applicant, including the expertise of his shareholders, suppliers, contractors and providers of engineering support who will be actively involved in the construction, marketing and operation of the cellular telephony service.

(f)   Full particulars of any financial interest the Applicant or its shareholders may have in other parts of the world in cellular communications or any other related communications business and should highlight their experience in this regard.

Description of Service

26.   (1)   Applicants must describe the services they intend to provide.

(2)   The information furnished on this aspect of the application will become an integral component of any licences issued.

(3)   The Minister also intends specifying in the licence the minimum annual coverage in terms of the population covered for each region as well as over the entire South Africa.

(4)   *This minimum coverage will be finally determined after the award of the licence and will be based on proposals contained in the applications of both C2 and C3.*

(5)   *An Applicant must therefore -*

(a)   *furnish particulars of his proposed annual coverage, properly identifying the areas concerned;*

(b)   *state that he guarantees compliance with such proposed coverage.*

Community Service Obligations

27. (1)   *Give details of general community services offered to all subscribers, either free or at a price.*

(2)   *Give detailed proposals on how penetration will be achieved even for subeconomic services and how the Applicant will implement and finance these services.*

(3)   *How will the Applicant market his services actively and at affordable prices to areas which are currently underprovided with telecommunications services.*

[Note : These services must be offered in both urban and rural areas. Services offered may entail the use of a different technology to that proposed for the mobile cellular networks.]

Market Analysis

28.   *The Applicant must project the size of the total market and how it is subdivided into different target groups. The expected number of subscribers should be given for the first 5 years for the different target groups.*

Planning and Project Management regarding Construction of Network

29.   *(1)*   *The Applicant must give full particulars of how he -*

    (a)   *intends to solve the planning and project management problems involved in setting up a cellular radio system in a relatively short period of time;*

    (b)   *intends to solve the problems of acquiring the necessary resources and equipment to set up and operate his cellular radio system;*

(c)   *proposes to acquire the necessary sites, other property, technology, personnel and capital.*

(2)   *The Applicant must furnish the projected costs of the infrastructure and express it as a percentage of the estimated total revenue to be earned during the term of the licence.*

System Design

30. (1)   *This part of the application requires information about the system design concept.*

(2)   *The technology to be employed is to be fully described. In this regard it must be confirmed that all equipment will comply with recognised international technical standards and specifications applicable to the required technology.*

(3)   *The recommendations of the ITU and their associated organisations (CCIR, IFRB) as they apply to South Africa must be observed.*

(4)   *Equipment must conform to the latest available test specification at the time of being taken into use.*

(5)   *Information on the approach to the network planning should be given later under "Planning Concepts". Under this item the following further information must be given :*

     (a)   *The technology being offered (eg GSM, TACS, NMT, AMPS, D-AMPS etc.) and the basic description of the system and network architecture including broad frequency requirements.*

     (b)   *Particulars of all interfaces in the network including details of the air interface. Specify whether the interfaces are standard or proprietary.*

     (c)   *Particulars of handover and roaming both within the Applicants' network and, if applicable, the alternate cellular network - i.e. within C2 and, if applicable, C3.*

     (d)   *Particulars of the proposed routing between an MS of one licensee and other subscribers who may be connected to*

-    *the fixed network (e.g. the PSTN)*
-    *the mobile communications network at another centre but of the same operator*
-    *a data network*
-    *the alternate cellular network*
-    *the international networks.*

(e)   The upgradability of the system with regard to new standards and technical developments.

Planning Concepts

31.   The following information must be supplied under this heading :

(a)   The basic approach to network development planning and how the system will be able to expand.   The limits to the system and the incremental cost effects must be stated.

(b)   Full details of the radio network planning, including -

    (i)   the exact approach used in the frequency plan

    (ii)   planning instruments to be employed inter alia

- forecast of propagation attenuations (inside and outside buildings) for the various BTS's
- interference analysis
- the antennae locations
- the methods used to preserve frequency economy;

(iii)   *adherence to EMC specifications;*

(iv)   *a guarantee that the equipment will not produce radio frequency interference which adversely effects sensitive electronic equipment such as hearing aids, heart pacemakers, video recorders etc. or pose a health hazard to any user;*

(v)   *quality targets.*

(c)   *Full details of fixed network planning including -*

(i)   *the exact approach used*

(ii)   *planning instruments to be employed*

(iii)   *key influencing factors*

(iv)   *theoretical traffic considerations*

(v)   *particulars concerning the efficient use of the fixed network of Telkom (transmission lines of Telkom, routing via the switched network, leased lines and gateways)*

(vi)   *numbering plan required*

(vii)   *alternate routing requirements.*

(d)   *Phases of network development (both the radio network and the fixed network planning) in at least the following 3 phases -*

    (i)   *initial phase*

    (ii)   *development and operational stages*

    (iii)   *ultimate capacity.*

(e)   *Particulars of the extent to which a faster a roll-out of service could be achieved by means of the infrastructure sharing with the other operator, especially in the non-urban areas? The relevant areas must be identified.*

    [Note : A plan for the temporary sharing of infrastructure will not be regarded as anti-competitive provided that quality of service and tariffs are not adversely affected.]

(f)   *Presentation of data from network development planning in the forms of tables, diagrams, maps, etc. for the initial phase, for at least two intermediate phases and for projected ultimate capacity.*

(g)   Data on the following items must be given for each phase :

    (i)   forecast number of users and the volume of communications

    (ii)   type and number of base stations

    (iii)   antennae locations

    (iv)   frequency and radio channel requirements

    (v)   type and number of mobile switching centres

    (vi)   structure and number of transmission lines required

    (vii)   type and number of gateways to other telecommunication networks.

    (viii)   mobile and fixed radio coverage with regard to geographical areas and population.

(h)   Details on the following operational issues :

    (i)   quality criteria and grade of service

    (ii)   network capability including blocking rate

    (iii)   failure/safety aspects

    (iv)   network redundancy

    (v)   network and fault management

    (vi)   organisation of network management

    (vii)   status provision, fault detection

    (viii)   servicing and maintenance concept

    (ix)   data management and billing systems

    (x)   any other information which may support the application.

(i)   Applicants proposed arrangements to safeguard the privacy of telecommunications and data protection.

Business Planning

32.   (1)   Full particulars must be given of the Applicant's business planning for the term of the licence.

(2)   Full particulars must be given of the fundamental assumptions underpinning the business planning including topography, demographic factors, costs, tariffs, licence fees, exchange rates, economic projections, escalations, etc.

(3)   Special mention must be made of the parameters over which the Applicant has little or no influence.

(4)   It is also important to show the sensitivity to the subscriber's costs as a result of changes in such parameters (including Telkom tariffs). Using the assumptions stated in response to 32(2) above as a mean, show how a 10% variation in these assumptions will affect the cellular tariffs.

(5)   *The Applicant's business planning should be supported by projected -*

(a)   *balance sheets;*

(b)   *profit and loss accounts;*

(c)   *cashflow statements; and*

(d)   *discounted cashflow statements,*

drawn up in accordance with General Accepted Accounting Practice (GAAP) for the period 1994 - 1999.

(6)   *Particulars must also be given of -*

(a)   *the key financial policies of the Applicant e.g. dividend rate, debt equity etc.;*

(b)   *financing requirements of the Applicant and how they will be met during the licence period;*

(c)   *outside financing arrangements (e.g. loans) and what foreign exchange risks are created;*

(d)   *the source of all finance.*

[Note : With regard to the use of the transmission lines and the other telephone services of Telkom, business planning must be based on the same rates provided by Telkom to all prospective Applicants to permit comparisons to be made of all submissions. The Telkom rates should be regarded as planning guidelines until finally agreed upon after negotiation.

Marketing

33.   *(1)*   *The Applicant must describe -*

(a)   *what type of service he will provide for various target groups;*

(b)   *what broad marketing strategies will be pursued (e.g. publicity, service, billing, advertising, quality of service);*

(c)   *what distribution strategy will be deployed for mobile stations (e.g. service providers, distributors, agents, etc.);*

(d)   *how the market will be developed.*

[Note : A licence holder may, subject to the approval of the Regulator, enfranchise the services it is entitled to provide e.g. by enfranchising service providers, regional operators and identifiable operators in niché markets. The licence holder shall, however, remain ultimately responsible for compliance with his licence conditions.]

Financial Resources

34.   (1)   *Applications must give full particulars of the Applicant's financial resources and how the whole operation will be financed during the term of the licence.*

(2)   *Detailed information must be given of guarantees by shareholders, other associated enterprises, suppliers and credit institutions, guaranteeing -*

(a)   *the financial obligations of the Applicant; and*

(b)   *compliance by the Applicant of his licence conditions.*

(3)   *The amount of external capital (both equity and loans) available for this project must be clearly indicated.*

Tariffs

35. (1) The Applicant must give particulars of all tariffs both at the service provider (if applicable) and subscriber level for the first year of operation and then projected tariffs to the fifth year using the assumptions called for in clause 32.

(2) Detail at least the following rates and fees -

(a) connection fee;

(b) monthly subscription fee;

(c) standard call charges (in and out);

(d) mobile to international;

(e) off-peak;

(f) roaming;

(g) charges for additional services (service and cost must be specified);

(h) tariff differentiation.

Competitive aspects

36. (1) The competitive implications of implementing the technical and business planning must be described.

(2)   In particular applications must fully describe -

   (a)   the method of developing the market and the role of service providers (i.e. third party retailers);

   (b)   how mobile stations will be distributed and sold.

(3)   A substantiated unit cost of the mobile stations must be given together with an estimate of how costs are likely to move in the future.

Information about Licence

37.   Applicants must ensure that they have provided sufficient information to permit a licence to be issued should they be successful. They should therefore study the draft structure of the licence, Annexure B.

Other information to be given

38.   (1)   The Applicant must in so far as such information has not been given under the previous paragraphs, provide such information as may be relevant in respect to the selection criteria detailed in chapter 5 (e.g. criteria such as local added value, job creation) as well as any other information required in the previous chapters of this document.

(2)   Provide all relevant information in order to enable the Minister to determine what companies are to be regarded as associated companies of the Applicant and the materiality of such association.

Interconnection Information

39.   (1)   Specify what information is required from Telkom to enable interconnection with their networks.

(2)   Give full details of all frequency requirements.

Services to be Provided by Telkom

40.   Specify the quality and quantity of services to be provided by Telkom.

Identification of Subscribers

41.   Applications must give particulars of the method used to identify individual subscribers, as well as the capabilities of the system to deal effectively with stolen mobile stations and other fraudulent practices.

CHAPTER 5

SELECTION CRITERIA

Criteria to be Applied

42. *In assessing the information contained in the applications the extent of satisfying the criteria set out hereinafter shall be taken into account.*

Proposed Service

43. (1) *The quality and facilities of the proposed service.*

(2) *The extent to which the choice of technology will lead to high volumes worldwide and low unit costs and will foster sustained competition in the mobile cellular market.*

(3) *The extent to which the technology will support and foster trade and industry in South Africa.*

(4) *The quality, facilities, reliability and affordability of the service and the appropriateness of technology.*

Expertise and Transfer of Know-how

44.　*(1)*　　The expertise of the Applicant to construct, operate and maintain the system.

　　　(2)　　The extent to which expertise in both the technology and operation of a cellular network is transferred to South Africa.

　　　(3)　　The extent to which local research, development and manufacturing will play a role in the provision of equipment and software.

　　　(4)　　The training to be given to South African employees in technology.

　　　(5)　　The contribution towards training and development of small businesses.

Community Service Obligations

45.　*(1)*　　The extent to which the Applicant is prepared to offer community services in addition to essential services such as directory and emergency services.

　　　(2)　　In particular, how the proposed service, or some extension to it, will provide telephone services to the underdeveloped sections of the South

African community where services may have to be rendered at sub-economic tariffs.

(3)   This criterion is a key determinant and will weigh heavily in the evaluation.

Experience

46.   *The relevant experience of the Applicant, his employees, subcontractors, suppliers and technical support groups who will take part in constructing, operating and maintaining an efficient cellular system.*

Planning

47.   *The planning capability to set up the cellular system within a short time.*

Business Plan and Strategy

48.   *The Business Plan and Marketing Strategy to secure high penetration at affordable rates which offer value for money to subscribers.*

Competition/Consumer Benefits

49.   *The extent to which the Applicant offers a competitive service, in particular the cost of mobile stations, tariffs (e.g. cost per minute of basic services) etc.*

Local Added Value

50. (1) *The extent to which the Applicant will support South African industry and the South African telecommunications manufacturing industry in particular.*

   (2) *The extent to which the Applicant will be involved in counter trade.*

   (3) *The extent to which the Applicant will contribute to local research and development.*

   (4) *The local content of equipment to be used by the Applicant either directly or via counter trade to achieve the same minimum targets.*

   (5) *The extent to which the service will contribute to exporting opportunities.*

   [Note : C2 and C3 will be required to achieve the same minimum targets which will be agreed before the licences are issued to them.]

Job Creation

51. *The extent to which the application will create jobs in South Africa, both as a result of the generation of local added value and in the operation and marketing of the service.*

South African Control

52.   *The extent to which South African citizens ordinarily resident in South Africa are involved in the "ownership" and management of the Applicant.*

Foreign Equity

53.   *The extent of foreign equity financing.*

Impact on Foreign Exchange

54.   *The extent to which the service will contribute towards officially supported foreign export credit for capital goods and services and the long term impact it will have on foreign exchange.*

Financial Resources of the Applicant

55.   *(1)   The financial resources of the Applicant and the guarantees (financial or otherwise) given that an efficient service will be rendered during the term of the licence.*

   (2)   The willingness of the Applicant to invest sufficient capital to provide an efficient service during the term of the licence.

Coverage

56.   *The extent of coverage and rate at which it will be achieved.*

Utilisation of Existing Infrastructure

57.   *Full particulars must be given of the extent to which existing infrastructure will be utilised efficiently, thereby creating revenue for the providers of such infrastructure (e.g. revenue share to Telkom).*

2.

A Regulator will be appointed to regulate the telecommunications industry. He will be assisted by a separate panel of experts who will advise the Minister on policy issues. The Regulator will also make sure that a level playing field will be created for the different operators.

The State's main aim with cellular telephony is to offer a significant advantage to lesser developed communities. It will be the Regulator's task to see to it that this purpose is achieved.

In accordance with the Government's policy of free enterprise and competition, licences will initially be granted to two operators, and it will happen by way of tender enquiry. More details on this will be available shortly.

In order to enable Telkom to share in the rapid development of telecommunications technology, is was decided to grant Telkom a share in one licence. Due to its huge capital obligation as the sole supplier of the fixed network, and further obligations to expand this network, it was decided to limit Telkom's share in one licence to 50 percent.

The conditions of practice for Telkom and its partner(s) will be determined and will be exactly the same as those applying to the other succesful tenderer. There will be no differentiation between the participating parties.

The Regulator will make sure that control of both licences will be in South African hands.

Issued by
Elsa Krüger, Ministerial Liaison Officer
Tel (021) 45-7260

MEDIAVERKLARING DEUR DR PIET WELGEMOED, MINISTER VAN VERVOER EN VAN POS- EN TELEKOMMUNIKASIEWESE

INSTELLING VAN SELLULêRE TELEFONIE IN SUID-AFRIKA

EMBARGO: Streng 23h00, Sondag 14 Februarie 1993

Die Regering het besluit om lisensies aan twee operateurs toe te ken om sellulêre telefonie in Suid-Afrika te bedryf. Die tegnologiese ontwikkeling op hierdie gebied sal aangewend word om mededinging in die telekommunikasiebedryf te skep.

Sellulêre telefonie hou groot voordele vir die land in, waarvan seker die belangrikste is dat dit makliker is om te onderhou en te installeer in afgeleë gebiede. Dit is die goedkoopste en vinnigste manier om aan die meeste Suid-Afrikaners toegang tot 'n telefoon te kan bied.

Hierdie ontwikkeling kan vinniger telefoonpenetrasie, wat tans met die vaste netwerk-ontplooiing nie finansieel haalbaar en bekostigbaar is nie, aan die gemeenskap bied. Hierdie tegnologie en die verwagte mededinging wat daarmee gepaard sal gaan, sal koste afdwing tot voordeel van die verbruiker. Dit bring Suid-Afrika ook in pas met die res van die wêreld.

Twee lisensies vir mobiele operateurs sal toegeken word en die Staat beho die reg voor om, soos die mark ontwikkel, verdere lisensies - onderhewig aan die beskikbaarheid van frekwensies in die toekoms - beskikbaar te stel.

Hierdie stap is moontlik gemaak deurdat frekwensies in die 900 megahertz spektrum beskikbaar geraak het. Daar is besluit om 10 megahertz aan elke operateur beskikbaar te stel.

Die bestaande C450 mobiele stelsel sal nie verder uitgebrei word nie en Telkom sal die ekonomiese voortbestaan daarvan van tyd tot tyd in heroorweging neem.

2.

Daar word 'n Reguleerder aangestel om die regulering van die telekommunikasiebedryf te hanteer. Hy sal bygestaan word deur 'n afsonderlike paneel kundiges wat die Minister van advies kan bedien oor beleidsaangeleenthede. Die Reguleerder sal ook toesien dat 'n gelyke speelveld vir operateurs geskep word.

Die Staat se oogmerk is dat sellulêre telefonie as tegnologiese ontwikkeling ook 'n beduidende voordeel vir agtergeblewe gemeenskappe moet inhou. Dit sal deel van die Reguleerder se taak wees om toe te sien dat hierdie oogmerk bereik word.

In ooreenstemming met die Regering se beleid van vrye mededinging, sal lisensies aanvanklik aan twee operateurs toegestaan word. Dit sal geskied by wyse van 'n tendernavraag. Nader besonderhede hieroor sal binnekort beskikbaar wees.

Ten einde Telkom te laat deel in die snelle ontwikkeling in tegnologie rakende die telekommunikasiebedryf, is besluit om aan Telkom deelname in een lisensie te gee. Weens sy groot kapitaalverpligting as alleenverskaffer van die vaste telefoonnetwerk en verpligting om hierdie netwerk uit te brei, is voorts besluit om Telkom se aandeel tot 50 persent in die een lisensie te beperk.

Die voorwaardes van bedryf vir Telkom en sy vennoot (e) sal bepaal word en dit sal dieselfde wees as wat vir die suksesvolle tenderaar sal geld. Geen onderskeid sal tussen die onderskeie deelnemende partye getref word nie.

Die Reguleerder sal verseker dat beheer van beide lisensies in Suid-Afrikaanse hande gevestig sal word.

Uitgereik deur
Elsa Krüger, Ministeriële Skakelbeampte
(021) 45-7260

LICENCE STRUCTURE

1.   *Legal confirmation of the licence to the Licensee giving full description of licensee including shareholding, etc.*

2.   *Scope of License*

    2.1   *To operate a cellular telephone network*

    Full description of operation, cross referencing to application.

    2.2   *Radio licenses:*

    2.2.1 *Base Transceiver Stations (BTS's)*

    2.2.2 *Mobile Stations (MS).*

    2.2.3 *Microwave links for MSC - BTS (if applicable.*

    (All frequencies will be specified after award has been announced.)

    2.3   *Limitations and restrictions.*

3. *Conditions of the Licence*

    3.1 Compliance with technical specifications of approved technology and compliance to be maintained with the service offered in the Application.

    3.2 Electro-magnetic Interference.

    3.3 Community Service Obligations:

        3.3.1 General nature - e.g. emergency services, telephone directory, etc.

        3.3.2 Specific to underdeveloped people of South Africa.

    3.4 Interconnection with Telkom (as per Annexure C).

    3.5 Relationship with other networks.

    3.6 Implementation of offers made in application.

    3.7 Share structure as approved with limitations on transfer of shares.

    3.8 Coverage plan and schedule.

    3.9 Quality of service.

3.10　*Provision of service in extra-ordinary circumstances.*

  3.10.1　*Emergency calls.*

  3.10.2　*Essential services.*

3.11　*Privacy of communications and data.*

3.12　*Conditions of transferability of licences.*

4.　<u>*Licence Fees*</u>

4.1　*The basic licence fee shall comprise two parts namely-*

  4.1.2　*a once off amount of R100 000 000 (one hundred million Rand) as directed;*

4.2　*Radio Licence Fees payable to the Postmaster General annually:*

  4.2.1　*Basic fee of R5 000 000.*

  4.2.2　*A fee of R20 000 per 200 KHz channel.*

4.3   *Escalation: The above radio licence fees will remain constant until March 1995 whereafter the fees may be increased.*

4.4   *Connection fees and tariffs payable to Telkom SA. (To be negotiated after the award of the licence with C2 and C3 simultaneously negotiating with Telkom.)*

5.   *Commercial Conditions*

5.1   *Competition.*

5.2   *Tariff structures.*

5.3   *Right to sublicence, appoint service and value added providers.*

5.4   *Non discrimination except on volume, terms and level of service.*

6.   *Validity and duration*

6.1   *Date of issue.*

6.2   *Date of commercial operations.*

6.3   *Duration of licence: 15 years renewable. An application for renewal shall only be refused if-*

(a)   *the licensee failed to comply with the conditions of the licence or the provisions of the regulatory legislation; or*

(b)   *the service rendered by the licensee was not of the required standard;*

(c)   *the Regulator is not satisfied that the licensee will materially comply with the conditions of the licence or the regulatory legislation.*

7.   *Conditions of Cancellation of Amendment*

7.1   *State of Emergency, War, Force Majeure, etc.*

7.2   *Non-compliance with conditions of licence.*

7.3   *Bankruptcy or liquidation.*

7.4   *Unacceptable changes in share structure.*

7.5   *Unacceptable service.*

DEPARTMENT OF POSTS AND TELECOMMUNICATIONS
REPUBLIC OF SOUTH AFRICA

DEPARTEMENT VAN POS- EN TELEKOMMUNIKASIEWESE
REPUBLIEK VAN SUID-AFRIKA

APPLICATION FOR PRIVATE RADIOCOMMUNICATION LICENCE
AANSOEK OM PRIVAATRADIOKOMMUNIKASIE-LISENSIE

For office use/Vir kantoorgebruik	
File no./Lêerno.	CE no./HI-no.

NO PAYMENT TO ACCOMPANY THIS APPLICATION FORM.
GEEN BETALING MOET HIERDIE AANSOEKVORM VERGESEL NIE.

FOR OFFICE USE ONLY/SLEGS VIR KANTOORGEBRUIK

Data type / Datatipe
P R C N E W P

Radio type / Radiotipe

Regional code / Streekkode

Licence no. / Lisensieno.

Base map co-ordinates / Basiskaartkoördinate

Group code / Groepkode

Govt. code / Staatskode

Language preference / Taalvoorkeur

Date from/Datum vanaf Year/Jaar Month/Maand Day/Dag

Date to/Datum tot Year/Jaar Month/Maand Day/Dag

PLEASE COMPLETE IN BLOCK LETTERS/VUL ASSEBLIEF IN BLOKLETTERS IN

Name of business
Naam van besigheid

OR/OF

Surname
Van

Initials
Voorletters

Department or section
Departement of seksie

Title
Titel

Date of birth/Geboortedatum
Year/Jaar Month/Maand Day/Dag

Identity/Passport/Alien no.
Identiteit-/Paspoort-/Vreemdelingno.

*Type
Tipe

Building/Name of farm/Plot
Gebou/Plaasnaam/Plot

Office/Flat no.
Kantoor/Woonstel-no.

Nationality
Nasionaliteit

Place of birth
Geboorteplek

Street
Straat

Street no.
Straatno.

Suburb
Voorstad

City/Town
Stad/Dorp

Postcode
Poskode

Language preference
Taalvoorkeur

Postal address
Posadres

Postcode
Poskode

Stand no.
Standplaasno.

Home telephone no. (including trunk dialling code)
Woningtelefoonno. (met inbegrip van hooflynskakelkode)

If temporary licence, quote period
Indien tydelike lisensie, meld tydperk

Date from/Datum vanaf
Year/Jaar Month/Maand Day/Dag

Date to/Datum tot
Year/Jaar Month/Maand Day/Dag

Business telephone no. (including trunk dialling code)
Besigheidstelefoonno. (met inbegrip van hooflynskakelkode)

This relates to the Identity/Passport/Alien no. Insert either I (Identity), P (Passport) or A (Alien).
Dit het betrekking op die Identiteit-/Paspoort-/Vreemdelingno: Voeg in I (Identiteit), P (Paspoort) of A (Vreemdeling).—

† Insert E for English or A for Afrikaans.
Voeg E vir Engels of A vir Afrikaans in.

PT 524.—1987/88

T 122

S08(04A)SvS

ANNEXURE C

INTERCONNECTION

It is contemplated that interconnection between the networks of the C2 and C3 licensees with the fixed network of Telkom will be arranged by way of an agreement to be negotiated and concluded by Telkom and the C2 and C3 licensees. The terms and conditions of such agreement should be fair and reasonable, duly balancing the interests of all parties and promoting the efficient use of the respective networks.

Detailed principles governing interconnection will be issued shortly and after further consultation with Telkom. Applicants are also free to submit their comments, suggestions and requirements in this regard as soon as possible.

The interconnection fees and tariffs that Telkom intend charging will also be issued shortly and these are to be used in the business plans submitted in their applications.

The interconnection agreement will be subject to the approval of the Regulator.

ANSWERS TO QUERIES ON INVITATION TO APPLY FOR A LICENCE TO PROVIDE
A NATIONAL CELLULAR RADIO TELEPHONY SERVICE

GENERAL QUESTIONS

*Q: Could the role of the Regulator and the Regulatory Framework under which he will
work, be clarified?*

*A: Until a new Telecommunications Act is in position, the Postmaster General will make
recommendations to the Minister (defined in paragraph 1(v) of the "Invitation to
Apply") in connection with telecommunication issues as governed by the Post Office
Act, 1958, as amended, and at this stage in particular with regard to issues
pertaining to the provision of a national cellular radio telephony service.*

END OF GENERAL QUESTIONS

SPECIFIC QUESTIONS

Important Note to Recipients

Q: Could the assurance be given that the Minister is serious about deregulating the telecommunications industry and does not intend to abandon the process of issuing cellular licences?

A: Applicants are assured that it is the serious intention to persevere with the deregulation of the telecommunications environment by issuing cellular telephony licences and allowing the provision of certain links in the cellular environment to be deregulated.

Q: Could the process to be followed regarding the issuing of licence be clarified?

A: The Postmaster General and his panel of experts will make recommendations to the Minister regarding the award of the licence.

APPENDIX 1

Paragraph 1 (f)

Q: *Could copies of the GSM Recommendations and applicable ETSI standards be provided?*

A: *As the type of system is not prescribed by the Postmaster General, Applicants must obtain and propose recognised international cellular standards themselves.   It is envisaged to endorse type approval of recognised international standards.*

Paragraph 1 (p)

See Question and Answer for Paragraph 1 (f).

Paragraph 1 (z)

Q: *Could the definitions of "net revenue", "income", "returns" and "funds collected on behalf of others" be further clarified?*

A: *Net revenue is the audited sales revenue, determined in accordance with General Accepted Accounting Practice (GAAP), on which the 5% ongoing annual licence fee will be levied.   The other concepts are similarly determined by GAAP.*

Q: *Are interconnect payments deductible from net revenue?*

A: *For the purposes of the licence applications, interconnect payments should be regarded as not deductible from net revenue.*

Paragraph 1(ee)

Q: *Does "service provider" also mean an entity offering analog cellular service as a retailer?*

A: *"Service provider" includes an entity offering analog cellular service as a retailer. Also refer to "Amendments".*

Paragraph 6(3)

Q: *Could the closing date for the submission of licence applications please be confirmed?*

A: *The closing date for all applications (ie C2 as well as C3) is at 12 noon on 30 June 1993, as stated in a press release by the Minister of Transport and of Posts and Telecommunications on 5 May 1993.*

Paragraph 6(5)

Q: *Could an assurance be given that the perceived differences in handling the C2 and C3 applications does not constitute a lack of a level playing field and an obstacle in the process of deregulating the telecommunications monopoly?*

A: *All applications will be subjected to the same process. Every attempt will be made to deal with applications in a reasonable and impartial manner to ensure that a fair and significant degree of deregulation is achieved in the cellular environment.*

Paragraph 6(6)

Q: *Why must both C2 and C3 switch on their systems on the same date?*

A: *Potentially C2 could switch on sooner than C3 and keep the latter out, and vice versa. In order to ensure a fair start, the start date will be specified at the time of licence for both licensees. However, should one operator not meet the switch on date, no change in the date will be permitted.*

Paragraph 6(7)

    See Question and Answer for paragraph 6(6)

APPENDIX 1

<u>*Paragraph 7*</u>

Q: *In view of a perceived preference for the GSM standard, would the task to extend telephone services to as many people as possible be clarified before an appropriate technology standard is selected?*

A: *The invitation allows Applicants to offer and motivate different technologies of their choice. The issue referred to is one of a number of critical factors which will all receive due consideration during the adjudication process and the licensing of both C2 and C3.*

Q: *Could it be clarified whether GSM is specified for the C2 and C3 systems, and why a comparison of other technologies with GSM is called for?*

A: *The invitation does not specify any specific technology for C2 or C3. It is anticipated that Applicants may offer other mature or upcoming systems. GSM is a recent development which is being promoted and adopted in many countries world-wide. A comparison with GSM by Applicants offering other technologies may assist the evaluation process.*

Q: *Could evaluation penalties and weighting factors for each cellular technology be*

specified?

A: *No evaluation penalties and weighting factors for cellular technologies have been*

specified thus far. Should any be specified, these will be subject to the discretion of

the Minister. Applicants are to describe the merits of the systems they propose.

Paragraph 8

Q: *Will C2 also be subjected to the selection procedure referred to in this paragraph?*

A: *Refer to the Answer for paragraph 6(5). C2 will be granted a licence, but the details*

of that licence will be specified after their application has been approved.

Paragraph 10(1)(a)

Q: *Could "South African citizen ordinarily resident in South Africa" please be defined?*

A: *Reference to the term "South African citizen" has no political connotation. It is our*

serious intention to ensure that a reliable, suitable and affordable cellular telephone

system be deployed in South Africa. A South African citizen ordinarily resident in

South Africa in this context means a person who lives in South Africa and conducts

his usual or main business activities in South Africa, or intends to do so should he

obtain a licence. The intention is to ensure that South Africans are given the first

right to exploit the business opportunities being developed in South Africa.

Paragraph 10(2)

Q: Are entities in which the Government of South Africa holds any interest disqualified on the basis of the Government having an interest in Telkom?

A: At this stage the Government is the sole shareholder of Telkom and Telkom as the present monopoly holder in telecommunications is allowed no more than a 50 % share-holding in a company to be formed by Telkom and other shareholders and which is assured of the C2 licence. The C2 application will also be subjected to the adjudication process together with applications for the C3 licence.

Paragraph 11(1)(b)

Q: In view of the extent of the licences being applied for, can the limitation of 200 pages be exceeded?

A: _This limitation may be exceeded by what the Applicant regards as reasonable to motivate his application._ In order to keep the evaluation period as short as possible, Applicants should exercise serious restraint concerning the length of their applications. Pages should be A4, single sided and single spaced.

Paragraph 11(1)(c)

Q: *May the limitation on the summary be exceeded?*

A: *The limitation of 25 pages for the summary <u>may not be exceeded.</u>*

Paragraph 11(2)

Q: *Does "alternative technologies" refer only to technologies different from GSM?*

A: *The Applicant has the freedom to offer more than one technology of his choice. "Alternative technologies" should be interpreted in this sense.*

Paragraph 11(3)

Q: *With reference to the introduction of other subjects, can this be done on additional pages to those indicated in paragraph 11(2)?*

A: *See the answer to paragraph 11(1)(b).*

Paragraph 15(1)

     See answer to paragraph 6(3).

Paragraph 17(1)

    See answer to paragraph 6(3).

Paragraph 17(2)

Q: *With reference to the date of 30 September 1993 until when the application must be irrevocably open, could this be made conditional on satisfactory progress being made in discussions between the selection of a C3 licensee and the issuing of a licence?*

A: *The significance of the date is that the application in the form submitted may not be withdrawn before 30 September 1993. If no award has been made by 30 September 1993, the application will remain valid until withdrawn by the Applicant. After an award has been made, the issuing of the licence will depend on negotiations on further conditions, if any, being concluded successfully. If negotiations are not successful, the Applicant may withdraw or the Postmaster General may cancel the award and make another award.*

Paragraph 18(1)

Q: *Does "comments" in this paragraph mean comments to the Postmaster General, the Regulator and his panel of experts?*

A: *"Comments" in this paragraph should be read as comments to the Postmaster General and his panel of experts.*

Q: *What are your intentions with regard to confidentiality?*

A: *All who have given notice of their intention to submit an application will be provided with the questions and the answers given. Further clarification which may result from comments received, will also be provided. The identity of Applicants making enquiries or submitting comments will not be disclosed. Only where a clear case is encountered which could compromise confidential commercial information, will the answer be dealt with discreetly.*

Paragraph 18(2)

See Question and Answer for paragraph 18(1)

Paragraph 18(3)

See Question and Answer for paragraph 18(1,

Paragraph 23(2)

Q: *If the licence is suspended or cancelled, will the obligation for payment of fees be removed?*

A: *The obligation for payment of the basic licence fee is governed by the provisions of the Post Office Act, 1958, as amended. The Postmaster General may determine a pro rata repayment from the date of suspension or cancellation, respectively. Fees payable under the Radio Act, 1953, as amended, will not be repayable.*

Paragraph 23(5)

Q: *Will direct interconnection between C2 and C3 be allowed?*

A: *Interconnection between MSCs of C2 and C3 is envisaged via Telkom's fixed network. This offers a possible choice of intermediate switching via the Telkom exchange network or direct connection by means of (n) x 2Mbps leased transmission links, which must be rented from Telkom.*

Q: *Would C2 and C3 be allowed to connect their networks directly to the Telkom international exchange?*

A: *Both C2 and C3 will be allowed to connect directly to the Telkom international exchange using (n) x 2Mbps leased links rented from Telkom.*

Paragraph 24

Q: *Will service providers be expected to contribute to licence fees?*

A: *The cellular network operator (licence holder) will be accountable for the payment of all licence fees referred to in this paragraph.*

Q: Will zero rating be considered for VAT for standard, emergency, community and sub-economic services?

A: For the purposes of submitting licence applications, emergency, community and sub-economic services should not be regarded as being exempted from VAT. Clarification will be sought when the business plans of the Applicants are available.

<u>Paragraph 24(1)(a)</u>

Q: Why is such a high licence fee called for?

A: The frequency spectrum is a national asset to be allocated to a licensee to use for commercial purposes for the duration of the licence. The amount was decided by the Government. The effect of the licence fee on service charges will be studied from the applications during the adjudication process.

<u>Paragraph 24(1)(b)</u>

Q: How was the 5% ongoing annual licence fee on net revenue determined?

A: The annual licence fee of 5 % on net revenue was a decision by the Government.

Q: Does the ongoing annual licence fee apply to revenue derived from profitable as well as non-profitable or marginally profitable areas?

A: For the purposes of the licence applications, the 5 % ongoing annual licence fee shall apply to income derived from all areas.

Paragraph 24(3)(b)

Q: Will the charge per channel be effective on granting or when the frequency is used?

A: The charge per channel will be effective as of date of assignment to the licensee.

Q: Is the annual licence fee of R20000 payable for each 200kHz channel duplex pair?

A: The annual licence fee is based on R20000 for each 200kHz channel duplex pair (ie 200kHz base transmit paired with 200kHz mobile transmit). For cellular networks using a different channel bandwidth, the licence fee per channel will be calculated in proportion to the bandwidth occupied.

ЛП ГЛГ\ПЛ Л

Paragraph 26(3)

Q: *Could the service areas envisaged and related population numbers and demographic*
      data be provided?

A: *Information pertaining to the South African population, e.g. numbers and geographic*
      distribution, can be obtained from the Central Statistical Service. Telkom can be
      approached for information on recorded demand for telephone services. These can
      be expressed in the form of working lines and waiting applicants per telephone
      exchange or, alternatively, per 1000 population in metropolitan, urban and rural
      areas. The Applicant is required to propose service areas and coverage himself.

Paragraph 27(3)

      Refer to the question and answer for paragraph 26(3)

Paragraph 27(3)

Q: *Does this paragraph mean that any alternative technology (including fixed wire) may*
      be proposed, or must such an alternative technology be a cellular technology
      implemented within the frequency spectrum indicated in paragraph 5?

A: *No constraints will be imposed on alternative technologies for the provision of*
      affordable services in urban and rural areas under-provided with telecommunication
      services. The primary objective here is to allow Applicants the option of freedom of

choice to supplement and extend their cellular networks by the most cost effective

means to offer at least basic forms of services to these areas as soon as possible.

Technologies based on other frequency bands may also be proposed, but their

acceptability will obviously depend on the availability of such frequencies. For the

purpose of preparing the licence applications, information on under-provided areas

can be obtained from Telkom - see Answer for paragraph 26(3). These areas will be

finally described in the licence conditions at the time when licences are awarded.

Paragraph 30(3)

Q: *In order to ensure that participants are working to the same international standards*
     as the regulatory body, could a list of standards be submitted to the regulatory body
     for confirmation?

A: *Applicants should quote the standards and specifications they intend using and*
     indicate to what extent these are stable.

Q: *Could recommendations of the ITU and their associated organisations as they apply*
     to South Africa and as they are applied under South African operations be provided?

A: *South Africa falls in Region 1 in terms of the ITU regulations. Applicants must obtain*
     the relevant recommendations directly from the ITU and associated organisations.
     Telkom and other fixed network providers (should these be licensed) may be
     approached directly for the technical interface requirements of the fixed networks.

Paragraph 30(4)

Q: *Could the in-force date for test specifications be clarified?*

A: *See Answer to 30(3). Applicants must obtain test specifications from manufacturers in accordance with the most recent status of the systems offered and submit these with their applications.*

Paragraph 30(5)(c)

Q: *Should it be assumed that the C2 and C3 licences are based on the same standard?*

A: *It should not be assumed that C2 and C3 are necessarily of the same standard. The Applicant must indicate how roaming would take place within his own system, and between his own system and the other system if it is assumed that the other system is of a compatible standard.*

Paragraph 31(b)(iv)

Q: *Could the form in which the guarantee is required, be clarified?*

A: *With regard to interference with sensitive electronic equipment, the guarantee can be in the form of proof of compliance with relevant international standards and national standards in countries where the systems have been deployed. With regard to health hazards to the user, the Applicant may describe the current status of his system in terms of standards and practices followed and submit an undertaking as to his approach towards improving the networks proposed should the need arise to introduce further measures, onwards and retrospectively, in the future.*

Paragraph 31(c)(v)

Q: *Will C2 and C3 be allowed to lease lines (e.g. 2 Mbps) from Telkom and use these to carry traffic directly between MSCs around the PSTN?*

A: *Yes. Also refer to the answer to paragraph 23 (5). Interconnection via Telkom's leased lines to convey cellular traffic between MSCs within C2 and within C3 and between MSCs of C2 and C3 will be allowed.*

Paragraph 32(1)

See Question and Answer to paragraph 32(5).

Paragraph 32(5)

Q: *With reference to the different time scales implied by paragraphs 32(1) and 32(5), could it be confirmed that 5 year figures will be adequate to evaluate the business case?*

A: *It is confirmed that the business case can be evaluated over a period of 5 years.*

APPENDIX 1

Paragraph 32(5)(d)

Q: *What discount rate should be applied, and what methodology to determine the terminal value as at 1999 should be adopted, to ensure consistency between the applications?*

A: *Applicants should choose what they consider appropriate.*

Paragraph 32 [Note]

Q: *When will Telkom's rates be available?*

A: *Telkom's rates are imminent and will be distributed immediately upon receipt.*

Paragraph 35

Q: *To what extent will Applicants be obliged to maintain tariffs detailed in their proposals when encountering the real competitive situation on switch-on of the networks?*

A: *It is envisaged that the Postmaster General must be informed of any proposed tariff changes before becoming effective. In a competitive situation, tariff changes must be advised at least 7 days in advance. In a non-competitive situation, tariff changes must be advised at least 30 days in advance and will be subject to the Postmaster General's approval. The Postmaster General's requirements will be finalised when the licences are issued.*

Q: What is the purpose of providing tariff information in the proposal?

A: The purpose of the tariff information called for is to compare the different applications and is therefore a material consideration. In particular, the sensitivity of the tariffs to factors outside of the Applicant's control must be given.

Paragraph 45(1)

Q: With regard to emergency services, will the Applicant be expected to provide such services, or is access only required?

A: Applicants are free to indicate what range of emergency services they can accommodate or provide. These services could include self-provided services, innovative services or access to already established services, etc.

Q: In the case of access only to emergency services, will these be specified or will the Applicant be able to choose to which emergency service to provide access?

A: It is envisaged that access to some emergency services may be specified.

Paragraph 53

Q: Could "the extent of foreign equity financing" be clarified?

A: Foreign equity capital is money moved into South Africa and employed as fixed capital in the company. The source is not significant. The Applicant must state what he intends doing as nothing is prescribed in the invitation. Movement of funds into and out of South Africa is subject to controls exercised by the South African Reserve Bank.

Q: Will South African citizens who do not reside in South Africa but who can bring 100% equity finance on their own account from abroad, be disqualified under paragraph 10(1)(a)?

A: No. Such a transfer of funds represents increased equity in South Africa and will be viewed in a positive light. Our efforts to develop cellular telephony to the advantage of all South Africans, including those who are disadvantaged, irrespective of race, colour, sex or religion are honest and *any reasonable application - with or without foreign equity - will receive impartial consideration.*

APPENDIX A

ANNEXURE B

Paragraph 2.2.3

Q: Does this paragraph suggest that C2 and C3 will be allowed to construct, operate and maintain their own microwave links between MSCs and Base Station Subsystems?

A: It is envisaged that the environment beyond the MSCs towards the Base Station Subsystems may be deregulated to some extent. Details will be released shortly. A primary objective will be to achieve cost effectiveness by allowing optimal use of all existing infrastructures to be made. Should the need arise, the Minister can permit both C2 and C3 to construct, operate and maintain their own microwave links in terms of powers granted by section 90A(1) of the Post Office Act, 1958, as amended.

Paragraph 4

Q: Why does Annexure B not refer to the annual ongoing licence fee?

A: The ongoing annual licence fee has been omitted in error. The following paragraph should be added:

     "4.1.3 an ongoing annual licence fee of 5 % of the net revenue of the licensee, payable within 14 days after the end of each licence year."

"ANSWERS"   22   Issued 1993-06-08

Q: *Will the basic licence fees and radio licence fees be tax deductible?*

A: *Assume for the business plan that licence fees are not tax deductible.*

Q: *Within which limits will the annual radio licence fees be constrained?*

A: *Escalation of the radio licence fees will be subject to the provisions of the Radio Act. For the purposes of preparing the business plans of the licence applications, however, the radio licence fees should be assumed as remaining fixed.*

Q: *Will the ongoing basic licence fee be a fixed percentage, and if not, within what limits will it be constrained?*

A: *For the purposes of preparing the business plans of the licence application the ongoing annual licence fee should be assumed as remaining a fixed percentage (i.e. 5%).*

Q: *Will the ongoing annual licence fee be levied before VAT?*

A: *For the submission of the licence applications, assume that the licence fee will be paid after VAT. The question could also be directed to income tax experts.*

END OF DOCUMENT

APPENDIX 2

AMENDMENTS TO THE DOCUMENT "INVITATION TO APPLY FOR A LICENCE TO PROVIDE A NATIONAL CELLULAR RADIO TELEPHONY SERVICE".

Paragraph 1

(w)   *"mobile cellular radio telephone services" and "mobile radio services" shall all mean bearer services, telecommunications services and supplementary services within the meaning of the GSM Recommendations.*

(cc)   *"roaming" shall mean the use of a mobile station in an area served by an MSC other than the home MSC operated by the same licence holder or the use of a mobile station on a mobile cellular telephone network other than the network to which a user subscribes;*

(ee)   *"service provider" shall mean a person (natural or corporate) offering mobile cellular radio telephone services as a retailer;*

(Editor's note: the word "digital" before cellular in the original document has been deleted)

Page 7

Editor's note: After paragraph 3, insert the following heading and paragraph:

"Regulator

3A. All references to the Regulator in the invitation document and its annexures shall, unless inconsistent with the context, be construed as references to the Postmaster General acting subject to the authority of the Minister."

Paragraph 6

Editor's Note: Add the following subparagraph:-

(8) The clear intention is to license a mobile cellular radio telephony service, but fixed network extensions and other technologies may be required, particularly to provide services in areas currently under-provided with telecommunication services - refer to section on Community Services Obligations. At the time of issuing the licence, the precise limitations (if any) of such extensions will be incorporated as licence conditions.

APPENDIX 2

<u>*Paragraph 23*</u>

(5)   *It is contemplated that the interconnection arrangement of the* <u>*MSCs of the cellular*</u> <u>*telephone*</u> *networks of the C2 and C3 licensees to Telkom's fixed networks will be arranged by way of an agreement to be negotiated and concluded by the C2 and C3 licensees and Telkom.*

<u>*Paragraph 27*</u>

(3)   *[Note: These services must be offered in both urban and rural areas. Services offered may entail the use of different technologies to that proposed for the mobile cellular networks.* <u>*It is envisaged that the C2 and C3 operators shall share the*</u> <u>*obligation to provide community services.*</u> <u>*They will therefore be allowed to*</u> <u>*supplement and extend their mobile cellular networks by whatever technology,*</u> <u>*including fixed wire, proves to be the most cost effective and the quickest to*</u> <u>*implement.*</u> <u>*The provision of links between the MSCs and BSCs and between the BSCs*</u> <u>*and the BTSs may be liberalised to some degree but does not exclude Telkom from*</u> <u>*providing links in these areas in competition with other infrastructure providers.*</u> <u>*Liberalisation of the mobile cellular business, including any liberalised*</u> <u>*interconnections within the cellular systems, will be effected by the Minister in terms*</u> <u>*of section 90A(1) of the Post Office Act, 1958, as amended.]*</u>

Paragraph 30

(3)   The recommendations of the ITU and their associated organisations (CCIR, IFRB, CCITT) as they apply to South Africa must be observed.

Paragraph 31

(g)(vii)   Type and number of gateways to Telkom's PSTN.

Paragraph 39

(1)   Specify what information is required from Telkom to enable interconnection with their networks. Also specify what information is required from other network providers, including Telkom, to provide any links to connect the various network equipment components of the licensee.

APPENDIX 2

<u>*ANNEXURE B*</u>

<u>*LICENSE STRUCTURE*</u>

<u>*Paragraph 2.*</u>   <u>*Scope of Licence*</u>

2.1   To operate a <u>mobile</u> cellular telephone network <u>with extensions as specified herein</u>
         <u>*under:-*</u>

<u>*Paragraph 4.*</u>   <u>*Licence Fees*</u>

<u>*Editor's Note*</u>*: Add the following paragraph omitted in error:*

     "4.1.3   an ongoing annual licence fee of 5 % of the net revenue of the licensee,
     payable within 14 days after the end of each licence year."

                                             <u>*END OF DOCUMENT*</u>

APPENDIX 3

TELKOM'S PROPOSED INTERCONNECTION CALL CHARGES

RATE	FOR DISTANCES UP TO AND INCLUDING (KM)	CHARGE PER MINUTE (CENTS)	
		(VAT EXCL)	(VAT INCL)
A	50	21	23.94
B	100	29	33.06
C	200	35	39.90
D	400	43	49.02
E	800	53	60.42
F	above 800	61	69.54

NOTES:
- For calls originated within the mobile network to the PSTN, distance will be measured from the cellular operators' MSC to the point of delivery within the PSTN.

- Volume discounts based on the undermentioned schedule will be allowed i.r.o. the annual call charges for usage of Telkom's PSTN:

  - above R10 million = 2,5% discount
  - above R20 million = 5% discount
  - above R30 million = 7,5% discount
  - above R40 million = 10% discount

- Calls made to foreign destinations and special services will be charged according to the call charges prescribed in the Tariff for Telecommunication Services.

TARIFFS FOR TELKOM'S MEGALINE LEASED LINE SERVICES

Type of Megaline Service	Radial distance (km)	PERMANENT SERVICES				TEMPORARY SERVICES			
		2 Mbit/s Excluding VAT		2 Mbit/s Including VAT		2 Mbit/s Excluding VAT		2 Mbit/s Including VAT	
		Installation Charge	Monthly Rental	Installation Charge	Monthly Rental	Installation Charge	Hourly Rental	Installation Charge	Hourly Rental
		R	R	R	R	R	R	R	R
Local Ends (within MRA)	Flat Rate	16 294,00	1 299,00	18 575,16	1 480,86	20 367,00	2,22	23 218,38	2,53
Local Ends across Exchange Boundaries	0 to 20	16 294,00	1 515,00	18 575,16	1 727,10	20 367,00	2,59	23 218,38	2,95
	>20 to 30	"	2 043,00	"	2 329,02	"	3,50	"	3,99
	>30 to 40	"	3 077,00	"	3 507,78	"	5,27	"	6,01
	>40 to 50	"	3 275,00	"	3 733,50	"	5,61	"	6,40
	>50 to 100	"	5 005,00	"	5 705,70	"	8,57	"	9,77
Main Link		Monthly Fixed Cost	Monthly Rental per km	Monthly Fixed Cost	Monthly Rental per km		Hourly Rental per km		Hourly Rental per km
	0 - 50	480,00	90,50	547,20	103,17	-	0,27	-	0,308
	> 50 - 200	3 104,76	37,99	3 539,43	43,31	-	0,15	-	0,171
	>200 - 400	7 201,11	17,51	8 209,27	19,96	-	0,10	-	0,114
	> 400	10 492,30	9,28	11 961,22	10,58	-	0,07	-	0,08

TARIFFS FOR TELKOM'S MEGALINE LEASED LINE SERVICES

Type of Megaline Service	Radial distance (km)	PERMANENT SERVICES 34 Mbit/s Excluding VAT — Installation Charge	PERMANENT SERVICES 34 Mbit/s Excluding VAT — Monthly Rental	PERMANENT SERVICES 34 Mbit/s Including VAT — Installation Charge	PERMANENT SERVICES 34 Mbit/s Including VAT — Monthly Rental	TEMPORARY SERVICES 34 Mbit/s Excluding VAT — Installation Charge	TEMPORARY SERVICES 34 Mbit/s Excluding VAT — Hourly Rental	TEMPORARY SERVICES 34 Mbit/s Including VAT — Installation Charge	TEMPORARY SERVICES 34 Mbit/s Including VAT — Hourly Rental
		R	R	R	R	R	R	R	R
Local Ends (within MRA)	Flat Rate	39 497,00	5 014,00	45 026,58	5 715,96	49 371,25	8,59	56 283,23	9,79
Local Ends across Exchange Boundaries	0 to 20	39 497,00	8 558,00	45 026,58	9 756,12	49 371,25	14,65	56 283,23	16,70
	>20 to 30	"	22 649,00	"	25 819,86	"	38,78	"	44,21
	>30 to 40	"	33 486,00	"	38 170,04	"	57,34	"	65,37
	>40 to 50	"	36 668,00	"	41 801,52	"	62,79	"	71,58
	>50 to 100	"	49 275,00	"	56 173,50	"	84,38	"	96,19
Main Link		Monthly Fixed Cost	Monthly Rental per km	Monthly Fixed Cost	Monthly Rental per km		Hourly Rental per km		Hourly Rental per km
	0 - 50	2 653,00	932,46	3 024,42	1 063,00	-	2,70	-	3,08
	> 50 - 200	23 130,51	522,90	26 368,78	596,11	-	1,75	-	2,00
	> 200 - 400	85 971,85	208,69	98 007,91	237,91	-	1,16	-	1,32
	> 400	144 415,02	62,58	164 633,12	71,34	-	0,83	-	0,95

TARIFFS FOR TELKOM'S
MEGALINE LEASED LINE SERVICES

Type of Megaline Service	Radial distance (km)	PERMANENT SERVICES				TEMPORARY SERVICES			
		140 Mbit/s Excluding VAT		140 Mbit/s Including VAT		140 Mbit/s Excluding VAT		140 Mbit/s Including VAT	
		Installation Charge	Monthly Rental	Installation Charge	Monthly Rental	Installation Charge	Hourly Rental	Installation Charge	Hourly Rental
		R	R	R	R	R	R	R	R
Local Ends (within MRA)	Flat Rate	65 906,00	13 786,00	75 132,84	15 716,04	82 382,50	23,61	93 916,05	26,92
Local Ends across Exchange Boundaries	0 to 20	65 906,00	20 814,00	75 132,84	23 727,96	82 382,50	35,64	93 916,05	40,63
	> 20 to 30	"	88 157,00	"	100498,98	"	150,95	"	172,08
	> 30 to 40	"	92 848,00	"	105846,72	"	158,99	"	181,25
	> 40 to 50	"	97 620,00	"	111286,80	"	167,16	"	190,56
	> 50 to 100	"	153353,00	"	174822,42	"	262,59	"	299,35
		Monthly Fixed Cost	Monthly Rental per km	Monthly Fixed Cost	Monthly Rental per km		Hourly Rental per km		Hourly Rental per km
Main Link	0 - 50	8 608,50	2 895,00	9 813,69	3 300,30	-	8,40	-	9,58
	> 50 - 200	59 561,00	1 875,86	67 899,54	2 138,48	-	5,96	-	6,79
	> 200 - 400	289313,00	727,09	329816,82	828,88	-	3,97	-	4,53
	> 400	495141,00	212,52	564460,74	242,27	-	2,84	-	3,24

PRINCIPLES AND PROPOSED CONDITIONS FOR INTERCONNECTION AND USE OF TELKOM'S TRANSMISSION PATHS, SUBJECT TO AMENDMENT UPON THE GRANTING OF A LICENCE

1.   **Background**

The principles listed below have been formulated in the context of the following considerations:

1.1   *Telkom has an obligation to provide a telecommunication transmission system throughout South Africa and therefore should enjoy the maximum opportunity to use its fixed network at a cost based tariff;*

1.2   *Telkom supports the local telecommunications manufacturing industry;*

1.3   *Any liberalisation of any part of the transmission network within the cellular radio telephony systems (C2 or C3) will be by specific authorization only of the Minister, and not by an indication of liberalisation of the transmission network;*

1.4   *The licences which will be issued to the two cellular operators will be granted by the Minster in terms of current legislation, i.e. Section 90A(1) of the Post Office Act of 1958, as amended. The licences will describe all aspects of the network, including*

Issued 1993-06-08

any rights granted by the Minister for operators to employ their own transmission links.

2.   **Conditions of Interconnection**

2.1   *Telkom is obliged to offer its fixed network (as defined in the "Invitation") for long distance transmission between MSC's, for interconnection of the MSC's to the PSTN and for the connection of elements within a cellular network to both cellular operators on precisely the same terms and conditions.*

2.2   *The cellular operators may not bypass the fixed network of Telkom (as defined) for MSC to MSC or MSC to PSTN connections unless Telkom is unable or unwilling to make the necessary transmission infrastructure available.*

2.3   *The use of leased lines from Telkom shall be permissible, including "far end" entry into the PSTN, and Telkom is obliged to offer such leased lines within a reasonable time unless it is unwilling or unable to do so, in which case the operators may make their own arrangements after due authorization by the Minister.*

2.4   *C2 and C3 shall be obliged to use the fixed network of Telkom for MSC to BSC or BSC to BTS where such services cross a property boundary, unless the operators can prove to the Minister that Telkom is unable to provide a satisfactory service in terms of:*

2.4.1   *Quality and grade of service.*

2.4.2   *Reliability.*

2.4.3   *Bandwidth.*

2.4.4   *Timing (provided that reasonable notice shall have been given.)*

2.4.5   *Price of Service.*

2.4.6   *Other reasons acceptable to the Minister.*

Should the operators wish to provide an alternative transmission link due to the unacceptability of Telkom's tariffs, they must prove to the satisfaction of the Minister that they are able to offer alternative lower cost service which is itself based on cost based tariffs. They must further show that their own service meets the same criteria of quality, reliability and bandwidth as that required from Telkom.

2.5   *Notwithstanding the foregoing Telkom shall not be obliged to offer any transmission service to either C2 or C3 for MSC to BSC or BSC to BTS links. Should Telkom decline to offer such service, then the operators will be authorised to install their own transmission links or be permitted to hire them from third party suppliers duly authorised to provide such links.*

Issued 1993-06-08

2.6   *Third party transmission operators will be licensed by the Minister to cater for the situations which may arise in terms of paragraphs 2.4 and 2.5 above on the following terms:*

    2.6.1   *Service shall be rendered in an arms length transaction.*

    2.6.2   *Any service offered shall be made available to both C2 and C3 on equal terms for equal service.*

    2.6.3   *Tariffs to be charged shall be subject to the same scrutiny by the Minister as the tariffs of Telkom.*

    2.6.4   *The transmission links shall only be used within the cellular network and for the purpose specified in the licence.*

2.7   *The final agreements between Telkom and C2 and C3, including tariffs, shall be negotiated after the award of the two licences.*

    In the negotiations leading to the signing of agreements between Telkom and the operators, C2 and C3 shall negotiate together on the one side and Telkom on the other.

Except for differences which are clearly acceptable to the parties due to differences between the networks of C2 and C3, the agreements are to be identical for both C2 and C3.

Whilst these agreements may be negotiated to the mutual satisfaction of all parties, they shall conform to the above-mentioned principles and furthermore, the agreements shall be subject to the approval of the Minister.

3. *Tariffs*

The proposed call charges governing interconnection to the PSTN and the tariffs for the use of leased lines on a "Megaline Service" basis for MSC to BSC and BSC to BTS connections are all contained in Appendices 3 and 4 and are to be used in the business cases of the operators. The final interconnection call charges applicable to the cellular operators will only be determined after the award of the licences.

4. *Conflict/Disputes*

Disputes shall be resolved by negotiation with the Minister (or his delegate) acting as arbitrator if necessary.

END OF DOCUMENT

▼ ▼ ▼

About the Author

Rachael E. Schwartz is Counsel with the law firm of Reid & Priest LLP, practicing international telecommunications and transactional law in Washington, D.C. She previously worked as a lawyer for several subsidiaries of Bell Atlantic Corporation: the Chesapeake & Potomac Telephone Company of Maryland; Bell Atlantic Mobile Systems, Inc.; and Bell Atlantic International, Inc., where she primarily was responsible for international joint ventures. Ms. Schwartz received her Juris Doctor in 1981 from the Georgetown University Law School and received Master of Laws degrees in corporate law and international law from the New York University School of Law in 1992 and 1995, respectively.

▼▼▼

INDEX

The Artech House Telecommunications Library

Vinton G. Cerf, Series Editor

Writing Disaster Recovery Plans for Telecommunications Networks and LANs,
   Leo A. Wrobel

X Window System User's Guide, Uday O. Pabrai

For further information on these and other Artech House titles, contact:

Artech House
685 Canton Street
Norwood, MA 02062
617-769-9750
Fax: 617-769-6334
Telex: 951-659
e-mail: artech@artech-house.com

Artech House
Portland House, Stag Place
London SW1E 5XA England
+44 (0) 171-973-8077
Fax: +44 (0) 171-630-0166
Telex: 951-659
e-mail: artech-uk@artech-house.com